获2024年度河南省重大科技专项（241100110200）资助

南瓜和辣椒
抗逆生理代谢及抗性基因挖掘

郭卫丽 著

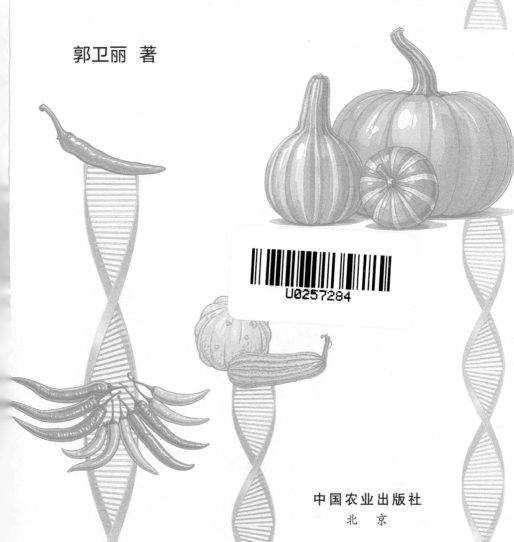

中国农业出版社

北 京

目 录

第1章 南瓜感染白粉病后的防御酶和光合特性变化 …… 1

1 文献综述 …… 1
  1.1 瓜类白粉病研究概况 …… 2
  1.2 抗病分子信号转导机制 …… 5
  1.3 研究目的与意义 …… 6

2 中国南瓜苗期性状与白粉病抗性的关系 …… 7
  2.1 试验材料 …… 7
  2.2 试验方法 …… 7
  2.3 项目测定 …… 8
  2.4 结果与分析 …… 9
  2.5 讨论与结论 …… 13

3 南瓜受白粉菌侵染后的活性氧暴发观察 …… 14
  3.1 试验材料 …… 15
  3.2 试验方法 …… 15
  3.3 结果与分析 …… 16
  3.4 讨论 …… 23

4 Photosynthetic properties and biochemical metabolism of *Cucurbita moschata* genotypes following infection with powdery mildew …… 25
  4.1 Introduction …… 25
  4.2 Materials and methods …… 27
  4.3 Results …… 30
  4.4 Discussion …… 35
  4.5 Conclusions …… 38

5 白粉病菌对结果期南瓜叶片光合特性和叶绿体超微结构的影响 ………………………………………………………… 39
  5.1 材料和方法 ……………………………………………… 39
  5.2 结果与分析 ……………………………………………… 42
  5.3 结论与讨论 ……………………………………………… 45

第2章 南瓜抗白粉病相关基因的分离及功能研究 …… 48

1 Improved powdery mildew resistance of transgenic tobacco overexpressing the *Cucurbita moschata CmSGT1* gene …… 48
  1.1 Introduction …………………………………………… 48
  1.2 Results ………………………………………………… 51
  1.3 Discussion …………………………………………… 62
  1.4 Materials and Methods ………………………………… 66

2 Expression of pumpkin *CmbHLH87* gene improves powdery mildew resistance in tobacco ……………………………… 72
  2.1 Introduction …………………………………………… 73
  2.2 Results ………………………………………………… 75
  2.3 Discussion …………………………………………… 87
  2.4 Materials and Methods ………………………………… 91

3 A pumpkin MYB1R1 transcription factor, CmMYB1, increased susceptibility to biotic stresses in transgenic tobacco …… 96
  3.1 Introduction …………………………………………… 97
  3.2 Materials and Methods ………………………………… 99
  3.3 Results ………………………………………………… 104
  3.4 Discussion …………………………………………… 112

4 A Pathogenesis-Related Protein 1 of *Cucurbita moschata* responds to powdery mildew infection ………………… 114
  4.1 Introduction …………………………………………… 114
  4.2 Materials and methods ………………………………… 116
  4.3 Results ………………………………………………… 122
  4.4 Discussion …………………………………………… 130

5 南瓜白粉病相关基因 *CmERF* 的克隆及功能分析 …………… 134
   5.1 材料与试剂 …………………………………………………… 135
   5.2 方法与步骤 …………………………………………………… 136
   5.3 结果与分析 …………………………………………………… 141
   5.4 讨论 …………………………………………………………… 147
   5.5 小结 …………………………………………………………… 150

第3章 辣椒对低温胁迫的响应与其低温抗性相关基因的克隆和功能分析 …………………………………………………… 151

1 文献综述 …………………………………………………………… 151
   1.1 茄科植物响应低温胁迫的研究进展 ………………………… 151
   1.2 逆境相关基因的研究进展 …………………………………… 157
   1.3 研究的主要内容 ……………………………………………… 164
2 外源 ABA 对低温下辣椒抗氧化酶活性及其基因表达的影响 ………………………………………………………………… 164
   2.1 材料与处理 …………………………………………………… 165
   2.2 试验方法 ……………………………………………………… 166
   2.3 结果与分析 …………………………………………………… 171
   2.4 讨论 …………………………………………………………… 177
3 SSH 技术对 ABA 调节的辣椒低温抗性相关基因的差异表达分析 ………………………………………………………………… 180
   3.1 材料与处理 …………………………………………………… 181
   3.2 试验方法 ……………………………………………………… 182
   3.3 结果与分析 …………………………………………………… 183
   3.4 讨论 …………………………………………………………… 196
   3.5 结论 …………………………………………………………… 201
4 辣椒 CaNAC2 转录因子的功能分析 …………………………… 202
   4.1 材料与方法 …………………………………………………… 203
   4.2 结果与分析 …………………………………………………… 212
   4.3 讨论 …………………………………………………………… 222
5 辣椒 CaMBF 转录辅激活因子的功能分析 …………………… 225
   5.1 材料与处理 …………………………………………………… 227

5.2 结果与分析………………………………………231
　　5.3 讨论……………………………………………240
6　利用 VIGS 技术研究辣椒 CaF-box 蛋白基因的功能………243
　　6.1 材料与处理………………………………………245
　　6.2 结果与分析………………………………………246
　　6.3 讨论……………………………………………253
7　全文结论及创新点……………………………………256
　　7.1 全文结论…………………………………………256
　　7.2 创新点……………………………………………257

参考文献……………………………………………………258

第1章 南瓜感染白粉病后的防御酶和光合特性变化

1 文献综述

南瓜是重要的蔬菜作物，我国主要种植中国南瓜（*Cucurbita moschata* Duchesn）、印度南瓜（*Cucurbita maxima*）和美洲南瓜（*Cucurbita pepo*）。南瓜为一年生藤本植物，栽培周期短，对环境适应性强，栽植广泛，中国大部分地区均有栽植。南瓜含有多种人体所需的营养物质（Wang et al.，2002），果实和种子中的多糖、类胡萝卜素和 β-胡萝卜素，具有提高机体免疫力、降低血糖、抗氧化等功能，在人体医疗和保健方面具有特殊功效（Yadav et al.，2010；Isutsa et al.，2013）。2011 年仅在中国，南瓜和西葫芦栽培面积 377 682 hm^2（约占世界栽培面积的 21.28%），总产量 6 978 167 t（占世界总产量的 28.71%）（Wu et al.，2014）。

南瓜白粉病是一种广泛存在的真菌性病害，在不同地区，不同白粉菌生理小种致病力有一定的差异，是一种危害南瓜生产的世界性病害。白粉病对环境的适应性强，可对南瓜进行周年侵染，目前用于生产的南瓜栽培种大多不具有白粉病抗性，而河南科技学院南瓜种质资源创新利用团队已挖掘出高抗白粉病材料 112-2 自交系。研究抗白粉病材料的抗病机制，利用分子生物学手段将抗白粉病基因引入到南瓜中，对南瓜的抗白粉病育种具有现实意义。

1.1 瓜类白粉病研究概况

1.1.1 瓜类白粉病发病条件

白粉病作为瓜类的重要真菌病害，近年来，通过对不同地区白粉病病原菌分生孢子的形态、萌发方式，菌丝的数量、颜色，闭囊壳的形成等方面进行鉴定，发现引起瓜类白粉病的病原均为子囊菌（*Sacfungi*）。引起我国南瓜白粉病的病原主要是单囊壳属白粉菌（*Podosphaera xanthii*, formerly *Sphaerotheca fuliginea*），且发病率高、传播速度快，在中国多个省份的南瓜主产区危害越发严重，成为限制南瓜生产的重要因素（梁巧兰等，2010；Zhou et al.，2010；咸丰，2012）。

瓜类白粉病的发生和流行与温度、湿度、光照度、pH、栽植密度、苗龄等条件有关（Braun，1987）。瓜类白粉病的适宜发病温度一般为20~25℃，南瓜白粉病在15~28℃范围内都可发病（梁巧兰等，2010），温度过高或过低都不利于白粉病的发生及传播。瓜类白粉菌不易形成闭囊壳，在20℃、相对湿度70%和4 400 lx的光强条件下产生，在其他温、湿度和光照条件下，形成分生孢子。白粉病一般在高温高湿和高温干旱交替时易发生，低温干燥条件下不易发生，湿度越大越易发生，但相对湿度过大也不易发生，光照弱或遮阴，粗放管理、种植密度大都利于白粉病的暴发（张焕春等，2014）。瓜类白粉病主要通过分生孢子侵染寄主，只能活体寄生，离体后很快失去活性。瓜类白粉病主要侵染叶片，老叶比嫩叶更易感病，发病初期叶面零星分布白色小斑点，随后白粉病病斑面积不断扩大，白粉病侵染部位叶片变黄，整个叶片覆盖白色霉斑。白粉病发病后期白粉病病斑由白色变为黄褐色，有老孢子或分生孢子闭囊壳产生，叶片发黄严重甚至焦枯，但不易脱落，严重影响植株正常光合作用，此外白粉病还可以侵染叶柄和茎部，但果实不易受到白粉菌的危害。白粉菌孢子接触叶片后，可在适宜条件下萌发，产生芽管和吸器从叶片表面侵入，进而产生菌丝，后期出现分生孢子梗，分生孢子可借助风力或者其他媒介传播到邻近的植株。白粉病发病后期可产生闭囊壳,闭囊壳在植物残体、发病植株和土壤中越冬,可随风或流水到达不同的地区,成为第二年的侵染源。

1.1.2 瓜类抗病育种

瓜类抗白粉病材料较少,露地和设施栽培中,白粉病成为危害瓜类生产的重要因素。目前,白粉病的防治主要依赖种子消毒、加强田间管理、药剂防治等措施(孟大山,2005)。但这些措施只能起到预防的作用,白粉病传播速度快,侵染寄主后4~5 d便可出现病症,进而危害整个植株。相关研究表明,抗白粉病南瓜材料比感白粉病南瓜材料产量高(Paris et al., 2002),选育抗白粉病材料可以有效解决白粉病对瓜类造成的危害。

(1) 白粉病抗性材料鉴定 近年来,南瓜白粉病越来越受人们的重视,对不同南瓜白粉病抗性鉴定的研究越来越多,筛选和鉴定种质资源的抗病性是开展抗病品种选育的基础。目前,主要通过大田和人工接种白粉病的方法进行抗白粉病鉴定,并具备白粉病病情指数分级标准(王建设等,2000)。南瓜抗白粉病材料的鉴定从苗龄上分为两种:幼苗期和成株期,在大田、温室或实验室对人工接种或自然感病的南瓜进行离体或者活体鉴定。离体接种是一种简便快速可靠的方法,可用于大量材料的鉴定,已经用于鉴定小麦、豌豆和葡萄的抗病性,在甜瓜上,离体叶片接种白粉菌也用于抗病性鉴定,且与活体接种获得的试验结果一致(邓丽君等,2015;王萍等,2010)。制备适当浓度的白粉菌孢子悬浮液喷雾法接种活体或离体叶片是一种简便快捷的方法。生理生化方面的研究取得了丰硕的成果,如POD、PAL、CAT、SOD、PPO及同工酶等酶活性的测定,其中部分已被判定为白粉病抗性指标。气孔是植物与外界环境进行气体和水分交换的重要通道,叶绿素是植物维持光合作用的重要物质,叶片表面的气孔数量越少,叶绿素含量越高,作物对白粉病的抗性越强(颜惠霞,2009)。

(2) 瓜类白粉病抗性遗传机制 目前,在瓜类作物与白粉病的互作研究中,有关黄瓜和甜瓜白粉病抗性遗传分析及其抗性基因连锁的分子标记报道较多（刘龙洲，2008；张海英等，2008；Wang et al.，2011；Kim et al.，2013；Ning et al.，2014）；但对瓜类作物抗白粉病基因分离与鉴定的研究有限，尤其在南瓜白粉病抗病基因的克隆方面十分欠缺。瓜类作物白粉病抗性遗传机制十分复杂，美洲南瓜对白粉病的抗性由不完全显性单基因（Pm-0）控制（Cohen et al.，2003）；蜜本南瓜（中国南瓜）白粉病的抗性是由单基因控制的隐性遗传（刘文君等，2013）。圆形中国南瓜白粉病 $C.$ $okeechobeensis$（small）Bailey 的抗性由 1 个不完全显性或显性的单基因控制（Contin et al.，1977）。

（3）瓜类抗白粉病研究进展　拟南芥三突变体（$AtMlo2$、$AtMlo6$ 和 $AtMlo12$）对白粉菌完全免疫（Consonni et al.，2006）。甜瓜 $CmMlo$ 家族基因中 $CmMlo2$ 与拟南芥 $AtMlo2$、$AtMlo6$ 和 $AtMlo12$ 高度同源，白粉病诱导其表达；$CmMlo2$-ihpRNAi 转化植株增强了拟南芥对白粉病的抗性，$AtMlo2$ 负调控植株对白粉病的抗性；但 $CmMlo1$ 和 $CmMlo3$ 受白粉病诱导下调表达（程鸿，2009）。高丽华从南瓜（$C.$ $moschata$ Duch.）中运用同源克隆法分离了 13 个 RGAs 片段，其编码的氨基酸序列均含有完整的 NBS 类抗病基因的保守结构域 P-loop、Kinase2、Kinase3a 和 GLPL（高丽华，2007）。通过构建黄瓜高抗白粉病品系叶片 cDNA 文库，发现一些 unigenes 参与植物抗病抗逆防御反应机制，其中 3 个 unigenes 与拟南芥白粉病抗性基因 $Mlo2$、$Mlo8$ 和 $Mlo14$ 同源性较高（齐晓花等，2010）。利用 SSH 技术分离和鉴定了黄瓜抗/感白粉病近等基因系（NIL210 和 D8）中的 DGEs，获得了 NIL210 中高表达的基因，包括抗病蛋白、泛素和 NBS-LRR 家族蛋白等广谱抗病蛋白，活性氧清除基因谷胱甘肽转移酶、苯丙氨酸裂解酶等以及信号转导分子丝氨酸/苏氨酸蛋白激酶等；并确定了抗病候选基因如 TIR-NBS-LRR 家族蛋白和苯丙氨酸裂解酶（Qi et al.，2010）。采用 cDNA-AFLP 技术分析甜瓜抗白粉病相关基因的差异表达模式，获得与抗白粉病相关的 25 条高同源 TDF，包括钙依赖钙调素不依赖蛋白激酶、ABC 转运蛋白 PDR8、水溶性环氧化合物水解酶和腺苷甲硫氨酸脱羧酶（咸丰，2012）。近年来，RNA-Seq 已被应用于瓜类的转录组分析，例如西瓜、黄瓜、甜瓜、苦瓜等（Guo et al.，2011；Ando et al.，2012；Blanca et al.，2012）。

1.2 抗病分子信号转导机制

植物抗病反应的信号转导途径有水杨酸（Salicylic acid，SA）途径、茉莉酸（Jasmonic acid，JA）和乙烯（Ethylene，ET）信号途径，这些途径能提高许多病原菌响应基因的表达。通过研究 SA、JA 和 ET 介导的信号传导途径的病程相关蛋白基因的表达模式，可以明确植物是通过哪些信号转导途径来调控抗病反应的。SA 信号传导的诱导因子有 R-Avr 互作、生物和非生物的刺激、专性寄生病原物的侵染等；JA 和 ET 信号通路主要受机械损伤、环境胁迫、病原物对根系的侵入、一些非病原细菌在根系繁殖等因子的诱导。不同的抗病信号通路有不同的信号传导调控因子及其激活的效应基因（Van Loon et al.，1998）。SA 介导的 SAR，主要针对专性寄生病原物，包括卵菌、病原细菌、真菌和病毒（Sticher et al.，1998）。植物被病原菌侵染后，SA 含量上升，某些病程相关蛋白基因的表达受 SA 的诱导。*NPR1* 基因位于 SA 信号转导途径的下游，发生突变后导致对病原菌的侵染敏感（乔禹等，2016）。SA 信号转导相关基因有 *NPR1*、*MYB6*、*PAL*、*PR1a*、*EDS1* 等；JA 信号转导途径的相关蛋白基因有 *CHIB*、*THI1.2*、*HEL* 和 *PDF1.2* 等；ET 信号转导相关的基因有 *ERF1*、*EIN2*、*ERF003*、*PR-3b* 等（郝欣，2016）。*PR1* 是拟南芥依赖 SA 的 SAR 抗病反应的一个标志基因，*PR5* 基因家族在南瓜、番茄、拟南芥、玉米等作物中都被发现，其编码的渗透素使病菌细胞质壁分离，抑制真菌和细菌生长发育（高婷婷，2013）。对拟南芥的研究表明 SA 信号转导途径主要参与对活体寄生病原物如霜霉病菌、假单孢杆菌、白粉病菌的防卫反应（Delaney et al.，1994；Liu et al.，2005），而 ET 和 JA 信号转导途径则主要对灰霉病菌、腐霉病菌和土传病菌等坏死营养型病原物的防御有调节作用（Thomma et al.，2000）。三种信号转导途径在参与抗病反应时不是相互独立的，而是相互影响的。

1.3 研究目的与意义

南瓜果实成熟的中后期和设施生产的坐果前期正值高温高湿环境，易暴发白粉病，且病害蔓延迅速，周年侵染，药剂防治效果往往不佳，抗病材料匮乏，极易造成叶片早衰减产；而且长期使用化学杀菌剂易使病原菌产生抗药性，同时给食用带来安全隐患，又加剧环境污染。因此，研究南瓜抗白粉病机制，挖掘抗病基因进行抗性材料培育是从根本上防治白粉病的有效途径。相对于黄瓜、甜瓜和西瓜，南瓜基因组公布较晚。目前对南瓜抗白粉病分子机制了解较少，关于南瓜抗白粉病相关基因的成功克隆和鉴定未见报道，研究主要集中在白粉病病原菌鉴定、生理生态和材料抗病性鉴定方面，以及少量报道的美洲南瓜（西葫芦）遗传图谱构建。利用抗白粉病的南瓜资源，了解南瓜与白粉病的互作关系，研究白粉病诱导下南瓜的差异表达基因，筛选白粉病抗性相关基因并对其进行功能鉴定，将是培育抗白粉病南瓜品种的有效途径之一。

研究的主要内容：对中国南瓜自交系 112-2 接种白粉菌 24 h 和 48 h 的转录组分析，发现 5 个与'九江轿顶'接菌处理表达量明显不同的 DEGs（*bHLH87*、*ERF*、*PR1*、*MYB1* 和 *SGT1*），采用同源克隆法克隆抗性候选基因。采用实时定量 PCR 技术分析南瓜白粉病相关基因在白粉菌和非生物胁迫下的表达模式；构建过量表达载体并用农杆菌介导的叶盘法遗传转化至烟草，获得转基因植株；分析转基因烟草对白粉病、青枯病和疮痂病的抗性，以及信号转导相关基因的表达模式。本试验利用南瓜高抗白粉病的珍贵种质资源挖掘抗病相关基因，进行烟草转化，验证其与白粉病抗性间的关系，为南瓜抗病分子育种提供有价值的基因资源。

2 中国南瓜苗期性状与白粉病抗性的关系

一般认为，表皮细胞的蜡质层、角质层的厚度，表皮细胞壁的结构（木栓化、木质化、钙化和硅化的程度），以及气孔、水孔等自然孔口的形状、大小、位置等都会影响寄主植物的抗病性（傅仕敏，2014；章珍等，2011；陈夕军等，2015）。在南瓜材料的选育中，育种家希望选出综合性状优良的单株或自交系，提出了综合分析法，以及改进的加权综合分析法等选择方法。然而，由于基因连锁与一因多效，考察的多个性状间通常存在一定的相关性（李新峥等，2009）。因此，我们推测南瓜的形态指标对白粉病的抗性有一定影响，现以中国南瓜自交系 112-2 和'九江轿顶'为试材，研究了不同材料南瓜的形态指标与白粉病抗性间的关系，以期为生产中选育优良的种质资源奠定基础。

2.1 试验材料

供试的中国南瓜自交系 112-2 和'九江轿顶'均取自河南科技学院。

2.2 试验方法

试验于 2015 年 10～11 月在河南科技学院园艺园林学院的温室内进行。10 月 1 日选种，挑选籽粒饱满、大小均匀一致的种子放于网袋中，并用铅笔注明材料的名称，每种材料选择 60 粒。于 10 月 2 日将种子放入温水（55℃ 左右）中进行温汤浸种，边搅拌边浸种 10 min，冷却后浸种 4～6 h。然后用湿毛巾包裹种子，并将其置于 28 ℃ 恒温培养箱中催芽。2015 年 10 月 4 日上午在温室内采用穴盘育苗，每穴播 1 粒种子，播种后覆上 0.5 cm 厚的细营养土，盖上薄膜，并将播种好的穴盘放入托盘内。

2.3 项目测定

2.3.1 形态指标测定

播种后结合温室内温度和土壤状况进行适量喷水，在白天温度过高时进行揭膜，以保证其通风透光。昼夜温差过大，晚上还应将膜覆上。喷水量以保证托盘内无积水为宜。待苗长到2叶1心期、3叶1心期和4叶1心期进行测量。每种南瓜材料选出长势整齐一致的10株，3次重复，分别测量其叶面积、叶长、叶宽、茎粗、茎长等形态指标，叶面积的测定方法采用方格法，对于叶宽、叶长和茎长用直尺进行测量，茎粗则采用游标卡尺测量。

2.3.2 叶绿素含量测定

采用丙酮法测定叶绿素含量（张志良等，2009）。

2.3.3 气孔密度测定

气孔密度的测量方法和计算方式（潘汝谦，2004）：将选好的新鲜真叶表面去污，用镊子撕去其上表皮制作临时玻片，放在显微镜上进行观测（用低倍镜还是高倍镜取决于表皮上气孔密度的数目），每个玻片取3个视野，求平均值，气孔密度单位为个/mm^2。

2.3.4 白粉病的病情指数统计

田间采集自然发病的南瓜叶片。用试管刷将病斑的白色霉层轻轻刷入无菌水中，制成接种液。采用喷雾接种法将接种液均匀喷到植株叶片上，以雾滴布满叶片但不滴为宜，随后将接种病菌的植株放入光照培养箱。苗长至一定大小进行白粉病的病情分析（刘淑艳等，2011）。根据下列公式计算病情指数：

病情指数＝$\sum$（病级代表值×该级发病叶片数）/（最高级代表值×调查总叶片数）×100

2.3.5 数据分析

形态指标、叶绿素含量、气孔密度以及病情指数均采用平均值±SD表示，采用SPSS软件对试验数据进行显著性和相关性分析。

2.4 结果与分析

2.4.1 两种南瓜材料的形态指标、叶绿素含量、气孔密度和病情指数分析

由表1-1可知，在2叶1心期南瓜自交系112-2叶绿素含量显著高于'九江轿顶'，其他性状间差异不明显。其中自交系112-2叶面积（63.60 cm^2）和叶长（6.50 cm）分别略低于'九江轿顶'的叶面积（72.40 cm^2）和叶长（7.82 cm）。

在3叶1心期自交系112-2和'九江轿顶'的形态指标差异不显著，自交系112-2茎长（6.84 cm）略高于'九江轿顶'茎长（5.30 cm）；而自交系112-2的叶面积（109.00 cm^2）、叶长（8.90 cm）和叶宽（12.20 cm）分别略低于'九江轿顶'的叶面积（115.40 cm^2）、叶长（9.54 cm）和叶宽（12.50 cm），而自交系112-2叶绿素含量（0.52 mg/g）高于'九江轿顶'叶绿素含量（0.44 mg/g），差异显著；自交系112-2气孔密度（381.34 个/mm^2）高于'九江轿顶'（346.88 个/mm^2），差异不显著；另外，自交系112-2病情指数（0.38）略低于'九江轿顶'（0.39），差异不显著。总之，自交系112-2和'九江轿顶'3叶1心期的形态指标、气孔密度和病情指数差异均不显著，叶绿素含量差异显著。

2种南瓜材料在4叶1心期形态指标(除茎粗外)存在显著差异，自交系112-2的叶面积（125.40 cm^2）、叶宽（9.76 cm）、叶长（9.56 cm），均低于'九江轿顶'叶面积（142.60 cm^2）、叶宽（13.64 cm）、叶长（12.30 cm）；自交系112-2叶绿素含量（0.96 mg/g）显著高于'九江轿顶'叶绿素含量（0.89 mg/g），自交系112-2气孔密度（489.56 个/mm^2）显著高于'九江轿顶'（416.37 个/mm^2），自交系112-2病情指数（0.39）显著低于'九江轿顶'（0.42）。总之，在4叶1心期，自交系112-2和'九江轿顶'的形态指标（除茎粗外）、叶绿素含量、气孔密度和病情指数均存在显著差异。

表 1-1 2 种南瓜材料不同时期的形态指标

时期	品种	茎长（cm）	茎粗（cm）	叶面积（cm²）	叶宽（cm）	叶长（cm）	叶绿素含量（mg/g）	气孔密度（个/mm²）	病情指数
2叶1心期	自交系112-2	4.24±1.94aA	0.48±0.016aA	63.60±38.25aA	9.80±1.78aA	6.50±1.95aA	0.48±0.03aA	314.77±79.45aA	0.30±0.01aA
	'九江矮顶'	4.22±1.89aA	0.47±0.016aA	72.40±42.68aA	9.64±2.93aA	7.82±1.42aA	0.39±0.05bA	289.91±47.59aA	0.38±0.01aA
3叶1心期	自交系112-2	6.84±1.60aA	0.59±0.04aA	109.00±35.20aA	12.20±1.20aA	8.90±1.10aA	0.52±0.02aA	381.34±59.46aA	0.38±0.01aA
	'九江矮顶'	5.30±1.40aA	0.59±0.04aA	115.40±27.20aA	12.50±1.30aA	9.54±1.20aA	0.44±0.02bA	346.88±46.38aA	0.39±0.01aA
4叶1心期	自交系112-2	7.50±1.60aA	0.68±0.06aA	125.40±39.50bA	9.76±1.60bA	9.56±1.60bA	0.96±0.01aA	489.56±60.32aA	0.39±0.01bB
	'九江矮顶'	6.28±1.40bA	0.59±0.04bA	142.60±43.20aA	13.64±2.50aA	12.30±2.00aA	0.89±0.02bA	416.37±75.43bA	0.42±0.01aA

注：同列数据后同小写字母表示差异显著（$P<0.05$），不同大写字母表示差异极显著（$P<0.01$）。

2.4.2 2种南瓜材料各形态指标、叶绿素含量、气孔密度和病情指数相关性分析

由表1-2可知，在2叶1心期南瓜材料（自交系112-2和'九江轿顶'）与叶绿素含量存在极显著正相关（相关系数为0.861）；叶绿素含量与病情指数呈极显著负相关（相关系数为－0.158）。叶面积与叶宽、叶长存在极显著正相关性（相关系数分别为0.971、0.881）。叶长与病情指数呈极显著负相关（相关系数为－0.781）。

表1-2 南瓜2叶1心期的形态指标、病情指数、气孔密度、叶绿素含量相关性

	材料	茎粗	茎长	叶面积	叶宽	叶长	病情指数	叶绿素含量
茎粗	0.046							
茎长	－0.487	0.400						
叶面积	0.143	0.418	0.521					
叶宽	0.114	0.567	0.525	0.971**				
叶长	0.271	0.051	0.211	0.881**	0.782**			
病情指数	－0.137	－0.037	－0.163	－0.627	－0.504	－0.781**		
叶绿素含量	0.861**	0.322	－0.443	0.297	0.359	0.325	－0.158**	
气孔密度	－0.224	0.018	－0.056	－0.447	－0.503	－0.458	－0.030	－0.436

注：**在0.01水平（双侧）上极显著相关，*在0.05水平（双侧）上显著相关。表1-3、表1-4下同。

由表1-3可知，在3叶1心期自交系112-2和'九江轿顶'2种材料与叶绿素含量呈极显著正相关（相关系数0.861），与气孔密度呈负相关（相关系数为－0.224）；气孔密度与叶绿素含量呈负相关（相关系数为－0.233）；叶绿素含量与病情指数呈极显著负相关（相关系数为－0.165）；气孔密度与病情指数呈负相关（相关系数为－0.238）。另外，叶面积与叶宽存在极显著正相关（相关系数为0.939）。

表 1-3 南瓜 3 叶 1 心期的形态指标、病情指数、气孔密度、叶绿素含量相关性

材料	茎粗	茎长	叶面积	叶宽	叶长	病情指数	叶绿素含量
茎粗	0.105						
茎长	−0.522	0.305					
叶面积	0.174	0.390	0.455				
叶宽	0.104	0.774	0.515	0.939**			
叶长	0.174	0.012	0.176	0.855**	0.794**		
病情指数	−0.070	0.000	−0.110	−0.659*	−0.524	−0.707*	
叶绿素含量	0.861**	0.287	−0.479	0.321	0.345	0.273	−0.165**
气孔密度	−0.224	0.015	−0.276	−0.337	−0.522	−0.405	−0.238	−0.233

由表 1-4 可知，在 4 叶 1 心期自交系 112-2 和'九江轿顶'材料分别与叶绿素含量显著正相关（相关系数为 0.731），与叶面积、叶长、叶宽显著正相关（相关系数分别为 0.870、0.768、0.873），与病情指数（相关系数为 −0.928）和气孔密度（相关系数为 −0.881）显著负相关；病情指数与气孔密度显著负相关（相关系数为 −0.837），与各形态指标，叶绿素含量显著负相关，相关系数依次为茎粗（−0.655）、茎长（−0.676）、叶面积（−0.821）、叶长（−0.752）、叶绿素含量（−0.692）。

表 1-4 南瓜 4 叶 1 心期的形态指标、病情指数、气孔密度、叶绿素含量相关性

材料	茎粗	茎长	叶面积	叶宽	叶长	病情指数	叶绿素含量
茎粗	0.733*						
茎长	−0.490	0.343					
叶面积	0.870*	0.900*	0.530				
叶宽	0.873*	0.851**	0.508	0.985**			
叶长	0.768*	0.872**	0.517	0.967**	0.976**		
病情指数	−0.928*	−0.655*	−0.676*	−0.821**	−0.837**	−0.752*	

(续)

材料	茎粗	茎长	叶面积	叶宽	叶长	病情指数	叶绿素含量
叶绿素含量 0.731*	0.395	0.195	0.394	−0.583	0.316	−0.692*	
气孔密度 −0.881*	0.775*	−0.315	−0.779*	−0.751*	−0.646*	−0.837*	−0.583

2.5 讨论与结论

南瓜作为一年生蔓性草本植物，材料丰富，形态指标较多。该试验对 2 种中国南瓜材料在 2 叶 1 心期至 4 叶 1 心期的形态指标进行分析，结果表明，在 2 叶 1 心期和 3 叶 1 心期自交系 112-2 和'九江轿顶'形态指标差异不明显（表 1-1），其中形态指标（茎粗、茎长、叶面积和叶宽）与病情指数存在负相关关系（表 1-2 和表 1-3）；植株长至 4 叶 1 心期，材料间形态指标差异达到显著水平（表 1-1），病情指数分别与各形态指标呈显著负相关，如茎粗、茎长、叶面积和叶长（表 1-4）。关于南瓜幼苗期形态性状与抗白粉病之间的关系很少见报道。该试验只是对 2 种中国南瓜材料在不同时期的形态指标对白粉病的抗性进行了分析，对于其他南瓜材料对白粉病抗性的分析有待于进一步的研究。

该试验在 2 叶 1 心期至 4 叶 1 心期中国南瓜自交系 112-2 病情指数小于'九江轿顶'（表 1-1），而且在 2 叶 1 心期至 3 叶 1 心期，材料（自交系 112-2 和'九江轿顶'）与病情指数呈负相关，4 叶 1 心时相关性达到显著水平（表 1-2 至表 1-4），表明自交系 112-2 的抗白粉病能力强于'九江轿顶'。这与周俊国等（2011）的试验结果相同。自交系 112-2 叶绿素含量高于'九江轿顶'，且差异显著（表 1-1）；且 2 种材料间的叶绿素含量与病情指数存在显著负相关性（表 1-2 至表 1-4）。因此我们推测自交系 112-2 的抗白粉病能力强于'九江轿顶'，与叶绿素含量存在正相关关系。此研究结果与颜惠霞（2009）对南瓜白粉病抗病材料及抗病机理的研究和刘会宁等（2001）对有关不同材料葡萄对霜霉病的抗性与叶绿素含量关系的研究结果相一致。因此，在生产实践中可将叶绿素含量作为衡量南瓜材料抗白粉病的一个初步鉴定指标。

  气孔作为植物进行气体和水分交换的重要器官，在植物的生长发育过程中具有重要的生理功能，但它也同样是植物病原菌侵染植物的重要通道，如白粉病菌则主要是通过气孔来侵染寄主的。该试验中，在4叶1心期，自交系112-2气孔密度大于'九江轿顶'（表1-1），2种材料间的气孔密度与病情指数存在负相关关系（表1-2至表1-4），长至4叶1心期负相关达到显著水平。因此我们推测自交系112-2的抗白粉病能力强于'九江轿顶'，与气孔密度存在正相关关系。针对气孔密度的这项研究与韩正敏（1998）等对杨树气孔密度和大小与黑斑病抗性关系的研究结果相同；与颜惠霞（2009）对南瓜白粉病抗病材料及抗病机理的研究结果和谢文华等（1999）对丝瓜气孔密度和大小与霜霉病的抗性关系的研究结果不同，所以在生产中对于气孔密度的应用应该视具体材料而定。

3 南瓜受白粉菌侵染后的活性氧暴发观察

  白粉病是南瓜栽培期间发生最严重且较普遍的一个病害，尤其对温室和大棚中种植的瓜类作物危害严重（吴仁锋等，2013），给生产带来了很大的经济损失。近年来的研究发现，当病原菌侵染植物体时会引起叶片中活性氧的暴发；活性氧暴发被认为是过敏性的特征反应，也是植物对病原菌应答的最早期防卫反应之一（Averyanov，2009）。活性氧主要是需氧细胞在正常的生理代谢活动中产生的，包括超氧阴离子、羟基自由基、过氧化氢、单线态氧等（张梦如等，2014）。在植物正常的细胞代谢活动中，产生活性氧的途径有多种，活性氧（ROS）积累较多会影响细胞的正常生理代谢功能，但同时活性氧又被认为是一种信号，可以使细胞在逆境中调控相关基因的表达和病原防御机制（崔慧萍等，2017；Casano et al.，2001）。植物与病原菌非亲和互作可以产生过敏反应（hypersensitive response，HR），寄主细胞受到病原菌侵染后会迅速坏死，以延缓或抑制病原菌的进一步生长和扩散（王教敏等，2009）。有研究指出，过敏反应和植物的活性氧暴发有密切关系，活性氧积累的水平和植物过敏性细胞的数量呈正相关（李征等，2006；柳利龙等，2014；李佩芳，2013）。李建武（2010）指出，黄瓜在与霜霉病的亲和互作中产生了大量活性氧，随后诱导发生过敏性反应，限制了病害的发生。同时活性氧可以调控植物相关基因的表达，引发自身的防御反应，参与系统获得性抗性的建立（薛鑫等，2013；陈年来等，2011）。当前有关白粉病侵染南瓜方面的研究主要集中在防御酶活性方面，在活性氧暴发水平和材料抗病性之间联系的相关研究仍然较少，本研究对白粉病菌侵染南瓜过程中活性氧的暴发情况进行观察，以探究活性氧和南瓜抗病性之间的关系，同时为进一步研究南瓜抗白粉病防御机制提供理论依据。

3.1 试验材料

3.1.1 白粉菌

采集河南新乡种植基地大棚中，表现白粉病的南瓜叶片。

3.1.2 南瓜品种

中国南瓜'九江轿顶'和中国南瓜自交系112-2，种子由河南科技学院提供。

3.2 试验方法

在培养皿中倒入少量无菌水，将南瓜叶片上的白粉病病斑霉层用软毛刷轻轻刷入培养皿中，随后用震荡仪震荡均匀，制成临时玻片，用显微镜观察孢子的数量，选用25×16计数板，共观察3个玻片，每个玻片选取5个视野，并按照下列公式进行计算。

$$白粉病菌细胞数 = \frac{80 小格内白粉菌细胞个数}{80} \times 400 \times 10^4 \times 稀释倍数$$

最终孢子悬浮液浓度为1×10^5个/mL。植株生长到三叶幼苗期和第1雌花开花结果期，将白粉菌孢子悬浮液喷施于植株叶片上。将制好的孢子悬浮液灌入喷壶中摇匀，于上午 9：00 开始喷菌，'九江轿顶'和中国南瓜自交系 112-2 各喷 21 棵，2 个材料各留 7 棵不进行喷菌，作为试验对照组。设置 0、24、48、72、96、120、144 h 7 个时间段检测活性氧（ROS）的爆发。

3.2.1　DAB 组织染色法检测 H_2O_2

植物叶片渗透二氨基联苯胺（DAB）来检测 H_2O_2 是一种较常见的组织染色方法。其原理是 H_2O_2 在过氧化物酶的催化下可与 DAB 迅速反应生成红棕色物质，从而定位叶片中的 H_2O_2（祁艳等，2007；徐晓晖等，2007）。剪取感染白粉病的南瓜叶片约 10 mm^2，浸泡在准备好的 DAB（1 mg/mL）染色溶液中。利用循环真空水泵在 0.1 MPa 下抽气、放气，直到叶片完全浸没在染色液中。随后将染色液浸泡的叶片在 35 ℃ 水浴锅中反应 4 h。取出叶片浸没在 75% 酒精中，70 ℃ 水浴加热 15 min 后更换酒精，重复水浴加热除去叶绿素，以上过程重复 3 次。用倒置荧光显微镜观察，叶片显现红棕色斑点，在可见光条件下拍摄照片（王敏，2011）。

3.2.2　氮蓝四唑（NBT）组织染色法检测 O_2^-

NBT 组织染色法检测 O_2^- 的依据：NBT 在 O_2^- 的作用下被还原生成不溶于水的蓝色物质，可原位显现 O_2^- 的存在（林金明等，2002；李响等，2013）。剪取的南瓜叶片约 10 mm^2，浸泡在准备好的 NBT 染色液中。利用循环水真空泵在 0.1 MPa 下抽气、放气，直至叶片浸没在染色液中。随后在 35 ℃ 水浴锅中反应 4 h。取出叶片然后浸没在 75% 酒精中，70 ℃ 水浴加热 15 min 后更换酒精，重复水浴加热除去叶绿素，以上过程重复 3 次。用倒置荧光显微镜观察，叶片显现蓝色斑点，在可见光条件下拍摄照片（赵晓玉等，2014）。

3.3　结果与分析

3.3.1　南瓜接种白粉菌后叶片表面观察

由图 1-1 可见，对照组中'九江轿顶'和中国南瓜自交系 112-2

第1章 南瓜感染白粉病后的防御酶和光合特性变化

幼苗期和结果期均没有观察到白粉病病斑；侵染24~96 h时，中国南瓜自交系112-2和'九江轿顶'均未观察到病斑；侵染120 h时，'九江轿顶'叶片观察到白粉，中国南瓜自交系112-2仍没有观察到病斑，说明中国南瓜自交系112-2比'九江轿顶'的抗性强，白粉菌侵染速度较慢。侵染24~120 h期间，中国南瓜自交系112-2和'九江轿顶'结果期均未观察到病斑。从'九江轿顶'不同生长阶段来看，幼苗期在侵染120 h观察到白粉病病斑，结果期未观察到白粉病病斑，说明南瓜结果期相较于幼苗期抗性强。

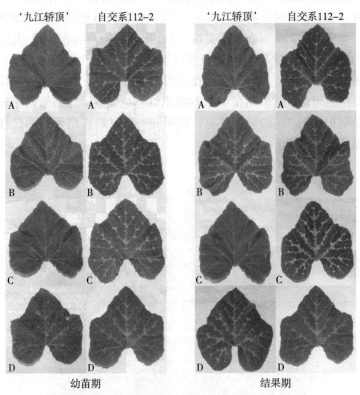

图1-1 不同南瓜材料幼苗期和结果期接种白粉病菌后叶片发病症状
A. 空白对照 B. 24 h叶片 C. 96 h叶片 D. 120 h叶片

3.3.2 南瓜接种白粉菌后 H_2O_2 的观察

在幼苗期（图1-2），自交系112-2和'九江轿顶'0 h接种上白粉菌孢子；24、48 h均没有棕红色斑点；72 h开始观察到少量棕红色斑点；96、120、144 h红棕色斑点继续增加，面积扩大连成片，颜色加深。自交系112-2接种72～144 h期间红棕色斑点较多，面积较大，表明自交系112-2接种后积累 H_2O_2 较快且含量较多。在结果期（图1-3），接种48 h自交系112-2叶片开始出现红棕色斑点，即开始产生 H_2O_2，72、96 h棕色斑点增加，120、144 h棕色斑点大面积增加，逐渐连接成片，而感病材料'九江轿顶'的植株72 h才观察到棕色斑点，96～144 h棕色斑点继续增多，颜色加深。综上所述，中国南瓜自交系112-2接种白粉菌后相较于'九江轿顶'材料出现的红棕色斑点数量多且颜色深，抗病性强的材料接种白粉菌后 H_2O_2 含量积累较快。在幼苗期，自交系112-2在72 h检测出红棕色斑点，结果期在48 h检测出红棕色斑点，即结果期相较于幼苗期积累 H_2O_2 的时间要早，说明同一种材料在结果期的抗性强于幼苗期。

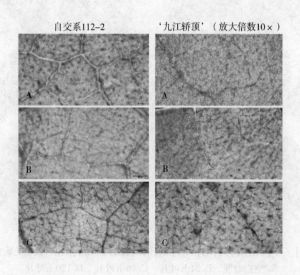

第1章 南瓜感染白粉病后的防御酶和光合特性变化

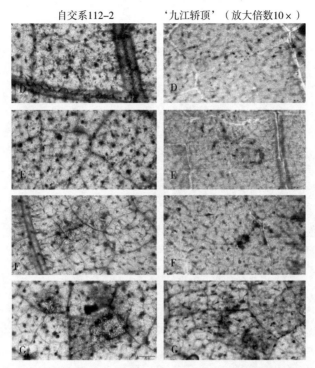

图 1-2 幼苗期不同南瓜材料接种白粉菌后 H_2O_2 积累量的观察
A. 0 h 叶片　B. 24 h 叶片　C. 48 h 叶片　D. 72 h 叶片
E. 96 h 叶片　F. 120 h 叶片　G. 144 h 叶片

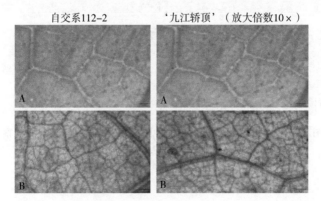

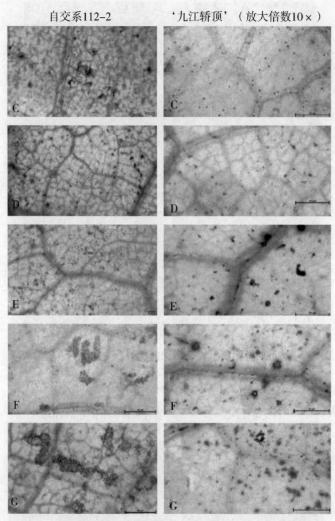

图 1-3 结果期不同南瓜材料接种白粉菌后 H_2O_2 积累量的观察
A. 0 h 叶片 B. 24 h 叶片 C. 48 h 叶片 D. 72 h 叶片
E. 96 h 叶片 F. 120 h 叶片 G. 144 h 叶片

3.3.3 南瓜接种白粉菌后 O^{2-} 的观察

在幼苗期（图1-4），中国南瓜自交系112-2和'九江轿顶'0 h接种白粉菌孢子；24 h开始出现蓝色斑点；48、72 h检测出蓝色斑点数量增加；96、120、144 h蓝色斑点继续增加，面积扩大，颜色加深。自交系112-2接种96~144 h期间蓝色斑点较多，面积较大，表明自交系112-2接种后积累 O^{2-} 较快且含量较多。在结果期（图1-5），中国南瓜自交系112-2与'九江轿顶'材料48 h开始观察到蓝色斑点，即检测到 O^{2-}，72 h中国南瓜自交系112-2蓝色斑点数量增加，开始连接成片，而'九江轿顶'材料只是蓝色斑点增加。综上所述，中国南瓜自交系112-2接种白粉菌后相较于'九江轿顶'材料出现的蓝色斑点数量多且颜色深，抗病性强的材料接种白粉菌后 O^{2-} 积累较快。在幼苗期，自交系112-2在24 h检测出蓝色斑点，结果期在48 h检测出蓝色斑点，即结果期相较于幼苗期积累 O^{2-} 的时间要晚。

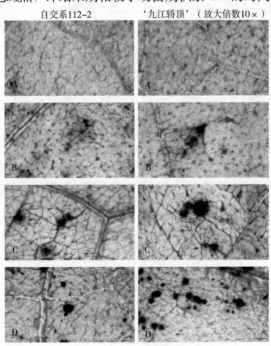

自交系112-2　　　　　'九江轿顶'（放大倍数10×）

自交系112-2　　'九江轿顶'（放大倍数10×）

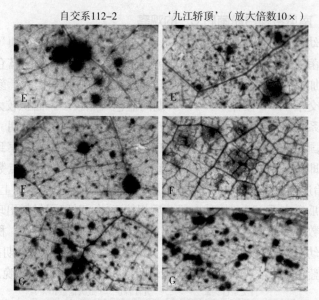

图1-4　幼苗期不同南瓜材料接种白粉菌后 O_2^- 积累量的观察
A. 0 h叶片　B. 24 h叶片　C. 48 h叶片　D. 72 h叶片
E. 96 h叶片　F. 120 h叶片　G. 144 h叶片

自交系112-2　　'九江轿顶'（放大倍数10×）

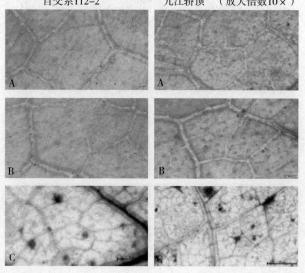

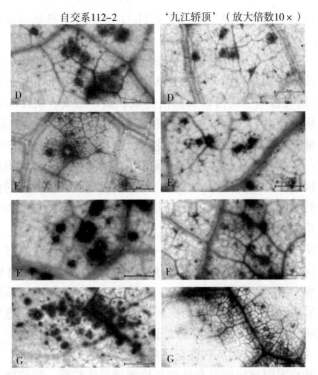

图 1-5 结果期不同南瓜材料接种白粉菌后 O_2^- 积累量的观察
A. 0 h 叶片　B. 24 h 叶片　C. 48 h 叶片　D. 72 h 叶片
E. 96 h 叶片　F. 120 h 叶片　G. 144 h 叶片

3.4 讨论

3.4.1 不同时期南瓜材料对白粉病的抗性鉴定

从南瓜不同生长阶段来看，侵染 120 h 后'九江轿顶'幼苗期（图 1-1）叶片出现了明显的白粉病斑，但结果期没有观察到白粉病斑，说明同一材料在结果期对白粉病的抗性要比幼苗期强。张素勤等（2008）研究表明，黄瓜苗期和成株期对白粉病抗性呈显著正相关，与本研究有相似的结论。

3.4.2 不同材料抗病性强弱与活性氧暴发间的关系

本研究发现，白粉病胁迫后，抗病材料中国南瓜自交系 112-2

和感病材料'九江轿顶'均产生了 H_2O_2 的积累，抗病材料积累的 H_2O_2 含量在幼苗期和结果期均高于感病材料，且抗病材料细胞积累 H_2O_2 较快、较多，感病材料细胞积累 H_2O_2 较慢、较少，说明不同南瓜材料的抗性强弱与 H_2O_2 的暴发量密切相关。裴冬丽等（2010）研究番茄感染白粉病的活性氧积累，结果表明在非亲和作用的抗病番茄细胞中活性氧积累更快，造成白粉菌无法正常生长；感病番茄在亲和作用下活性氧变化则恰恰相反。司盛伟（2011）研究发现，黄瓜叶片在接种霜霉病菌后，感病品种的 H_2O_2 积累量显著低于抗病品种。周威等（2007）研究白粉病菌入侵对不同抗性南瓜品种的病理和生理影响，结果表明 H_2O_2 的积累量与寄主的抗病性呈正相关，与本研究结论一致。

自交系112-2和'九江轿顶'在白粉病胁迫后均检测出 O^{2-} 的积累，抗病材料积累的 O^{2-} 较感病材料更多，说明南瓜材料抗病性强弱和 O^{2-} 的暴发量密切相关。王玲平（2001）研究黄瓜感染枯萎病菌后生理生化变化及其与抗病性关系得出，黄瓜叶片感染枯萎病后抗病品种的 O^{2-} 的积累量大于感病品种。

3.4.3 南瓜不同时期抗病性与活性氧暴发间的关系

从叶片表面感病情况和活性氧检测结果比较看，幼苗期'九江轿顶'在120 h观察到明显的白粉病斑，自交系112-2未观察到白粉病斑，但在120 h之前就能检测出活性氧；结果期2种材料均未观察到病斑，却同样检测出活性氧，说明在观察到明显的感病症状之前，植物已通过活性氧的积累来抵抗白粉菌的侵染。'九江轿顶'幼苗期（图1-2）在72 h检测出DAB染色斑点，结果期（图1-3）在48 h检测出DAB染色斑点，即结果期相较于幼苗期积累 H_2O_2 的时间要早，因此结果期抗性强可能与适量浓度的 H_2O_2 扮演信号分子介导抗病反应有关（汪承润等，2012；韩欢欢等，2012）；由图1-4和图1-5可知，两种材料在幼苗期24 h即检测出NBT染色斑点，结果期在48 h检测出NBT染色斑点，即结果期相较于幼苗期积累 O^{2-} 的时间要晚，说明结果期抗性强可能与结果期细胞对 O^{2-} 胁迫的敏感程度低于幼苗期有关（段学武等，2005）。

研究结果显示，白粉病胁迫后，细胞活性氧积累水平呈不断增加趋势，芦光新等（2016）研究指出，白粉菌初生侵染和次生侵染小麦细胞后的活性氧积累水平不同，在初生侵染 120 h 后，细胞活性氧积累水平降低，而本研究中在白粉菌侵染 120 h 后，仍能检测出大量活性氧，说明不同作物以及不同抗性材料对白粉菌侵染的耐受性不同，而次生侵染时南瓜细胞活性氧暴发量仍需进一步研究。

综上所述，通过叶片表面观察发现幼苗期'九江轿顶'在 120 h 观察到明显的白粉病斑，自交系 112-2 未观察到白粉病斑；而在结果期 2 个南瓜材料均未观察到白粉病斑，说明结果期抗白粉病能力强于幼苗期。在南瓜生长的幼苗期和结果期，自交系 112-2 在感染白粉病后观察到染色形成的红棕色斑点和蓝色斑点较多，而'九江轿顶'染色形成的斑点较少，说明自交系 112-2 在感染白粉病后相较于'九江轿顶'积累了更多的 H_2O_2 和 O^{2-} 来抵抗白粉病的生长。

4 Photosynthetic properties and biochemical metabolism of *Cucurbita moschata* genotypes following infection with powdery mildew

4.1 Introduction

The genus Cucurbita is composed of several species, including those that are cultivated *Cucurbita moschata* Duch., *Cucurbita pepo* L. and *Cucurbita maxima* Duch. and several wild species. *C. moschata* is an economically important species that is cultivated worldwide and has high productivity. Pumpkins are valued for their fruits and seeds and are rich in vitamins, amino acids, flavonoids, phenolics and carbohydrates. Pumpkins also have important medicinal properties, including antidiabetic, antioxidant, anticarcinogenic and anti-inflammatory activities (Yadav et al., 2010). Cucurbit powdery mildew (PM), mainly caused by *Podosphaera xanthii* (*P. xanthii*) (formerly *Sphaerotheca fuliginea*) (Perez-Garcia et al., 2009;

Fukino et al., 2013), is a serious biotrophic pathogen disease in field and greenhouse cucurbit crops worldwide. In pumpkin, PM mainly affects the leaves, with chlorosis and yellowish spots appearing in the early stages of the disease that gradually expand to spread white powdery matter across the whole leaves. PM substantially reduces pumpkin photosynthesis and causes premature desiccation of the leaves that together decrease the quality and marketability of the fruits.

Photosynthesis is a critical process for crop growth and performance, and it can be dramatically impaired when leaves are infected by foliar pathogens (Berger et al., 2007). Pathogen infection reduces the photosynthetic rate, which is usually associated with damage to the photosynthetic apparatus and increased excitation energy that exceeds that required for photosynthetic metabolism. These effects ultimately lead to the reduction of molecular oxygen and formation of reactive oxygen species (ROS); in particular, H_2O_2 (Bassanezi et al., 2002; Kumudini et al., 2008; Behr et al., 2010; Iqbal et al., 2012). ROS and H_2O_2 have fundamental roles in plant defense mechanisms, and act directly as antimicrobial/antiviral agents and indirectly by inducing the hypersensitive response (HR). This leads to the expression of β-1, 3-glucanases, and pathogenesis-related (PR) proteins that are encoded by systemic acquired resistance (SAR) genes (Seung and Hwang, 1996) and strengthening of the plant cell walls via cross-linking of glycoproteins (Jones and Dangl, 2006; Torres, 2010). Therefore, examining the leaf gas exchange parameters net photosynthetic rate (Pn), stomatal conductance to water vapor (G_s), transpiration rate (Tr) and internal CO_2 concentration (C_i) can be used to fully assess the photosynthetic performance of plants under pathogen infection. These properties can be used to provide insights into the mechanisms underlying plant-pathogen interactions (Rolfe and Scholes, 2010).

ROS production is an important plant defense strategy against pathogen infection (Magbanua et al., 2007). However, an imbalance between ROS production and removal can lead to ROS accumulation and damage to the host tissue (Lima et al., 2002; Scandalios, 2011). To maintain homeostatic control of ROS, plants have antioxidative enzymes and metabolites, including superoxide dismutase (SOD), which catalyzes the dismutation of O^{2-} to H_2O_2 and O_2; catalase (CAT), which dismutates H_2O_2 to O_2 and H_2O; and guaiacol peroxidase (POD), which reduces the H_2O_2 produced by SOD to H_2O and is the key enzyme involved in H_2O_2 scavenging. POD is activated by overproduction of ROS and is an oxido-reductive enzyme that participates in the reinforcement of the cell wall through suberization, lignification and cross-linking. It also activates defense genes in response to pathogens (Borden and Higgins, 2002). In addition, phenylalanine ammonia lyase (PAL) catalyzes the first committed step in the phenylpropanoid pathway, which is believed to play a critical role in regulating lignin accumulation in plants.

In this study we aimed to shed light on the mechanisms of resistance to PM. We compared the PM-susceptible "Jiujiangjiaoding" genotype (JJJD) with 112-2, a PM-resistant inbred line. For each genotype we measured the PM-infectivity, photosynthetic parameters and defense-related enzyme activities of both genotypes following PM pathogen inoculation.

4.2 Materials and methods

4.2.1 Plant materials and PM pathogen inoculation

Two genotypes of *C. moschata* were tested in this study, inbred line 112-2 and cultivar JJJD, which have been shown to be resistant and susceptible to PM, respectively. Seeds for each of these genotypes were provided by the Henan Institute of Science and Technology, Xinxiang, Henan, China (Zhou et al., 2010). Line 112-2 has been inbred through eight consecutive generations of self-pollination and exhibited high PM resistance in an 8-year outdoor field observation study. After at least 80% of the seeds had germinated, they were placed in 9-cm-deep plastic pots containing a 1 : 1 mixture of soil and peat in a growth chamber under long-day conditions (15 h of 5 500 lux light at 28℃ and 9 h dark at 18℃) to grow for 4 weeks before pathogen inoculation.

P. xanthii conidia were collected from naturally infected pumpkin leaves in a local greenhouse. A spore suspension at 10^6 spores per mL was made by soaking heavily infected leaves in tap water containing 0.01% Tween-20. When the seedlings had formed 2-3 fully expanded leaves (at approximately 4 weeks), the pots were divided into four groups. The four groups included the two control groups, named 112-2-CK and JJJD-CK, and the two PM treatment groups, named 112-2-PM and JJJD-PM. The control groups were sprayed with distilled water and the PM treatment groups were sprayed with the spore suspension at 08:00-09:00 hours until the surfaces of the seedlings was completely wet. Immediately after inoculation, the seedlings were transferred to a growth chamber as described. The control plants were kept in a separate growth chamber. The treatments were arranged in a randomized complete block design that consisted of four independent biological replicates.

4.2.2 Determination of severity of PM infection

At 6 days post inoculation (dpi), disease severity of the seedlings was scored on a 1 to 9 scale and calculated as a weighted average for each of these entries as described by Thomas et al. (2005). Circular disks of 2.5 cm diameter were excised from the young upper leaves of each seedling and placed in a sterile petri dish lined with moist blotting paper. The leaf disks was inoculated by adding 10 μL of the spore suspension (10^6 spores/mL) to the adaxial surface of the disk. Disease severity of the leaf disks was calculated as [($5A+4B+3C+2D+E$) /$5F$] ×100 according to Ishii et al. (2001).

4.2.3 Microscopic examination of PM pathogen infection

The third leaves (Three leaves from each treatment per replication) of inoculated and control seedlings of both genotypes were excised at 12, 24, 48, 72, 96 and 120 h of pathogen inoculation (hpi) and processed for microscopy observation Guo et al. (2018).

4.2.4 Photosynthetic parameters assay

Leaf gas exchange was measured using a Li-6400 Portable Photosynthesis System (Li-Cor, USA). The P_n, G_s, C_i and Tr were measured on the third leaf from 08:00-11:00 hours (solar time). The irradiance level was set at 600 μmol photons $m^{-2} s^{-1}$ and all measurements were performed at ambient temperature (25℃) at 1, 2, 4, 6, 8, 10, 12 and 14 dpi.

4.2.5 Antioxidant enzyme activity assays

Samples from the second and third leaves of four separate seedlings were collected at 0, 24, 36, 48, 72, 96 and 120 hpi. Leaf samples were collected from the control seedlings at each corresponding time point. The leaf samples were kept in liquid nitrogen during the sampling and then stored at −80℃ until required. The H_2O_2 content and the activities of SOD, CAT and POD were measured as described previously (Guo et al., 2012). The β-1,3-glucanase activity was measured by using laminarin as a substrate and assaying for D-glucose liberated by β-1,3-glucanase (Souza et al., 2017). The PAL activity was measured using by using L-phenylalanine as a substrate and assaying for the amount of trans-cinnamic acid (TCA) generated by PAL (El-Shora, 2002; Souza et al., 2017).

4.2.6 Statistical analysis

All data are expressed as the mean ± standard deviation (SD) of four independent biological replicates ($n = 4$). The data from replicates of the four treatments (112-2-PM, 112-2-CK, JJJD-PM and JJJD-CK) were subjected to two-way analysis of variance (ANOVA), and differences in the mean values between different treatments were determined using the least significant difference (LSD). Statistical procedures were performed using the statistical analysis system software DPS, version 7.55. Values for $P \leqslant 0.05$ were considered as statistically significant.

4.3 Results

4.3.1 Evaluation of PM resistance in *C. moschata* genotypes

The severity of disease in the two PM pathogen-inoculated genotypes was visually appraised (Table 1-5). The disease index of in vivo seedlings and in vitro leaf discs from the PM-susceptible JJJD genotype were 37.8 and 55.3, respectively. These values were significantly higher than those of the PM-resistant 112-2 genotype at 6 dpi. Fungal growth was microscopically assessed at 0-120 hpi (Fig. 1-6). No geminated *P. xanthii* conidia were detected for either of the genotypes at 0 hpi (Fig. 1-6a, g). On the leaf discs from the 112-2 seedlings, the *P. xanthii* conidia began to differentiate germtubes at 24 hpi, primary hyphae appeared at 48 hpi and the hyphae bifurcated to form secondary hyphae at 72 hpi (Fig. 1-6b ~ d). However, on the leaf discs from the JJJD seedlings, hyphae were first detected at 24 hpi and these bifurcated to form secondary hyphae at 48 hpi and a dense hyphal network at 72 hpi (Fig. 1-6h~j). At 96 hpi, a relatively high number of hyphae and conidia maturing in chains were observed on the JJJD leaf discs and these were obviously more abundant than those detected on the 112-2 leaf discs (Fig. 1-6e, k). At 120 hpi, the conidia development of the fungus of the JJJD leaf discs also became more mature and dolioform (Fig. 1-6f, l). Overall, the PM growth on the 112-2 genotype leaf discs was distinctly slower than that on the JJJD phenotype leaf discs as determined by both the visual estimates and microscopic observation. Together, these data suggest that there are differences in the responses to PM between the two genotypes that may be associated with different PM resistance mechanisms.

Table 1-5 Disease index of pumpkin seedlings inoculated with powdery mildew

Genotype	Disease index	
	in vivo seedlings	in vitro leaf discs
Inbred line 112-2	18.50 ± 1.23	26.67 ± 2.35
cv 'JJJD'	37.80 ± 2.41*	55.30 ± 3.52*

* indicates significant differences ($P<0.05$).

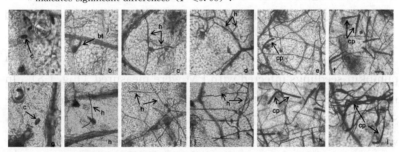

Fig. 1-6 Microscopic observation of *Podosphaera xanthii* infection in pumpkin seedlings

Powdery-mildew (PM) -inoculated leaves of genotype 112-2 at 0 h (a), 24 h (b), 48 h (c), 72 h (d), 96 h (e) and 120 h (f); and PM-inoculated leaves of genotype JJJD at 0 h (g), 24 h (h), 48 h (i), 72 h (j), 96 h (k) and 120 h (l). The arrow indicates growth of the PM pathogen; h, hypha; c, *P. xanthii* conidium; gt, germtube; cp, conidiophore.

4.3.2 Changes in the photosynthetic properties of pumpkin seedlings infected with PM

The P_n, G_s and Tr indices in both the 112-2-PM and JJJD-PM treatments were obviously reduced in comparison with the corresponding controls (Fig. 1-7a ~ c). These results suggest that PM pathogen inoculation reduces stomatal opening to limit PM pathogen invasion, which also leads to a reduction in the transpiration rate. The P_n in the 112-2-PM and JJJD-PM treatments were significantly reduced, by 51% and 76%, respectively, at 6 dpi in comparison with the controls. In contrast, the C_i in both the 112-2-PM and JJJD-PM treatments were increased in comparison with the controls (Fig. 1-7d). The above-mentioned indicators in leaves was not significantly different between both pumpkin genotypes infected with PM. The P_n of 112-2-PM was slightly higher than that of JJJD-PM after 4 dpi, and the G_s of 112-2-PM was higher than that of JJJD-PM at different time points.

4.3.3 Changes in the H_2O_2 contents and defense-related enzyme activities in pumpkin seedlings infected with PM

The H_2O_2 contents and defense-related enzyme activities in pumpkin seedlings infected with PM were measured (Fig. 1-8). In the 112-2-PM treatment, the H_2O_2 content (except at 48 h), SOD activity (except at 120 h) and CAT activity (except at 120 h) were reduced in comparison with the 112-2-CK control (Fig. 1-8a~c). The POD activity was increased in comparison with the control (Fig. 1-8d), suggesting that there is an insufficient ROS accumulation in the 112-2-PM seedlings to activate this ROS-scavenging antioxidant enzyme. The H_2O_2 content (except at 72 h) and the SOD activity in the JJJD-PM treatment were reduced in comparison with the JJJD-CK control, while the CAT and POD activities were higher than the control (except at 36 h). The

第1章 南瓜感染白粉病后的防御酶和光合特性变化

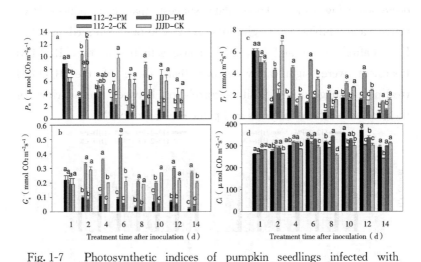

Fig. 1-7　Photosynthetic indices of pumpkin seedlings infected with *Podosphaera xanthii*

Net photosynthetic rate (P_n) (a), stomatal conductance (G_s) (b), CO_2 assimilation rate (C_i) (c), and transpiration rate (Tr) (d) in the leaves of control (CK) and powdery mildew-infected (PM) pumpkin seedlings. The values are the means ± SEs of four biological replicates. Data between treatments were analyzed by two-way ANOVA, and lowercase denotes statistical significance at $P \leqslant 0.05$.

SOD, CAT (except at 72 h) and POD activities in the 112-2-PM treatment were lower than those of the JJJD-PM treatment throughout the infection process, and the effect of interaction genotype×stress treatment on CAT and POD activities reached a significant level. This suggests that there is a negative correlation between antioxidant enzyme activities and plant resistance to PM. The PAL activities in both the 112-2-PM (except at 36 h) and JJJD-PM treatments were increased in comparison with their corresponding controls (Fig. 1-8e). While the β-1, 3-glucanase activities in both genotypes infected with PM were significantly decreased (Fig. 1-8f). These data suggest that PM induces PAL activity and inhibits β-1, 3-glucanase activity in pumpkin seedlings. The PAL (after 48 h) and β-1, 3-glucanase activities in the 112-2-PM treatment were significantly higher than those of the JJJD-PM treatment, suggesting a positive correlation between the activities of these defense-related enzymes and plant resistance to PM.

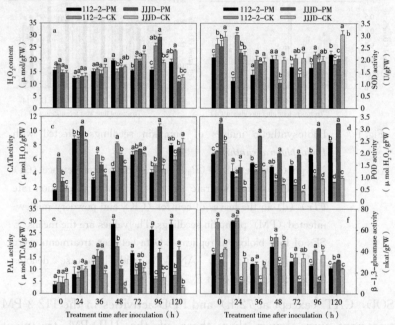

Fig. 1-8　H_2O_2 contents and defense-related enzyme activities in pumpkin seedlings infected with *Podosphaera xanthii*.

H_2O_2 content (a) and superoxide dismutase (SOD) (b), catalase (CAT)(c), peroxidase (POD)(d), phenylalanine ammonia lyase (PAL) (e) and β-1, 3-glucanase activities in the leaves of control (CK) (f) and powdery-mildew-inoculated (PM) pumpkin seedlings. The values are the means ± SEs of four biological replicates. Data between treatments were analyzed by two-way ANOVA, and lowercase denotes statistical significance at $P \leqslant 0.05$.

4.4 Discussion

Disease symptoms, including foliarspots with intense chlorosis and necrosis with low development of hyphae were detected at 24 hpi, while, maturation of *P. xanthii* conidia after 120 hpi shows that the whole infective process was completed during the 120 hpi. In contrast, the development of the fungal colonies was significantly reduced in the leaves of the PM-resistant 112-2 seedlings at 120 hpi. The results suggest that the changes observed in the growth of the PM pathogen are closely related to the differences in PM resistance between the two genotypes.

Photosynthesis is the most important metabolic process in plants, in which light energy is converted to chemical energy (ATP), reducing power (NADPH), metabolic intermediates and organic compounds (Scharte et al., 2005). Inducing a reduction in photosynthesis may be one of the primary strategies by which biotrophic pathogens establish a compatible interaction with their plant hosts (Bolton, 2009). The P_n was significantly reduced in the PM-inoculated seedlings, by 51% and 76% for the resistant and susceptible genotypes, respectively, in comparison with the controls. Moreover, the disease index at 6 dpi was 18.5 and 37.8 in the resistant and susceptible genotypes, respectively. Overall, the reductions in the P_n measured during the course of PM infection took place in parallel with reductions in the G_s and the Tr. The possible closure of stomata on leaves infected by PM may have contributed to these reductions. In addition, the C_i in both the 112-2-PM and JJJD-PM treatments were increased in comparison with their corresponding controls. This finding may be associated with the necrosis and chlorosis that often occurs as PM infection progresses, which abrogates the transport of water and solutes. This ultimately leads to stomatal closure and increases the C_i in the leaves of plants infected with PM and is associated with a reduction in the amount of green leaf tissue (Polanco et al., 2014; Tatagiba et al., 2016). The reduced photosynthesis in pathogen-infected leaves can occur due to changes in the opening and closure of stomata. This hinders CO_2 diffusion in the mesophyll and can lead to the reduction or destruction of chlorophylls or chloroplasts, which in turn results in chlorosis and necrosis of the leaf tissue (Rolfe and Scholes, 2010).

In this study, the H_2O_2 contents in both the 112-2-PM (except at 48 h) and JJJD-PM (except at 72 h) treatments were reduced in comparison with the control. The SOD and POD activities were reduced and increased, respectively, in the seedlings after inoculation with PM. These data suggest that the reduction in the H_2O_2 content following PM inoculation in these plants was insufficient to activate these antioxidant enzymes to scavenge excess ROS. Taking account into the low H_2O_2 concentrations detected, we hypothesize that the crucial role of POD in the PM defense mechanism is its participation in the reinforcement of the cell wall through lignification, not its role as a H_2O_2-scavenging antioxidant enzyme. This hypothesis is supported by a similar report that suggested that POD plays an important role in host defense; in particular, through host tissue lignification (Rauyaree et al., 2001). The activities of SOD, CAT and POD in the PM-resistant seedlings were lower than those in the PM-susceptible seedlings during the infection process. These findings suggest that the activities of antioxidant enzymes are negatively correlated with pumpkin resistance to PM. As expected, the reduced H_2O_2 content was associated with a decrease in the SOD activity, and this could not be attributed to the rapid removal of H_2O_2 by other disposal systems that minimize H_2O_2 cytotoxicity (Perl et al., 1993). Similar results were obtained in a previous study, which found that increased POD activity was correlated with resistance to downy mildew in cucumber(Ding et al., 2016). SOD and peroxidase activity have been associated with cucumber resistance to PM (Kang, 2009). Among the enzymes that are involved in removing excess H_2O_2, CAT plays a key role (Mittler, 2002). The reduced CAT activity in the PM-resistant genotype and the increased CAT activity in the PM-susceptible genotype may be associated with their different levels of resistance to PM. In general, it appears that reduced CAT activity increases host resistance to pathogen attack by maintaining a high H_2O_2 concentration (Magbanua et al., 2007).

PAL, a key enzyme of the general phenylpropanoid pathway, is involved directly or indirectly in the biosynthesis of phenolic compounds, phytoalexins, lignins and others secondary compounds. Increased PAL activity has been reported in plants that are resistant to virus infection (Siddique et al., 2014). These plants also have elevated levels of phenolics and lignin (Souza et al., 2017). Phenolic compounds are used by plants to generate the lignin precursors coumaryl, synapil and coniferyl. Plant cell wall lignification allows the formation of a physical barrier that prevents pathogens from spreading through the plant tissues (Ngadze et al., 2012). In our study, PAL activity was induced in both the PM-resistant and susceptible genotypes, but the PAL activity at 48 hpi in the 112-2-PM treatment was distinctly higher than that of the JJJD-PM treatment. This shows that the PM resistance of pumpkin genotypes is positively correlated with their PAL activity and suggests that this correlation might be associated with lignin accumulation.

Increased β-1, 3-glucanase activity in virus-infected plants favors susceptibility to viruses and cell-to-cell movement of viruses via plasmodesmata (Iglesias and Meins, 2000; Souza et al., 2017). Here, we found that inoculating 112-2 and JJJD with PM reduced β-1,3-glucanase activity, suggesting that reduced β-1,3-glucanase activity might contribute to PM resistance in pumpkin. Furthermore, PM grew more slowly in the PM-resistant 112-2 genotype than in the PM-susceptible JJJD phenotype, indicating that the higher POD and PAL activities in 112-2 are likely vital for limiting PM infection and systemic disease establishment in these plants.

4.5 Conclusions

Powdery mildew is one of the most common fungal diseases in *Cucurbits* based on recent studies, it severly affects the plant growth and reduces production. In our experiment we concluded that PM resistance of pumpkin seedlings is associated with the growth stage of the pathogen in parallel with the up/down-regulation of key enzymes in the antioxidant system. Increased POD and PAL activities are probably vital for the 112-2 PM-resistant genotype seedlings to restrict PM infection and systemic disease establishment. In contrast, the JJJD PM-susceptible genotype showed severe disease symptoms, fast pathogen growth, reductions in photosynthesis-related parameters and increased CAT activity; which together might have favored PM infection. Moreover, the PM infection alters stomatal closure, degrades mesophyll and chloroplast tissues, which leads to reduced photosynthesis in cucurbits leaves. Therefore, in future, more consise researchs should be carried out to understand the role of plant hormonal mechanisms and genetic expression in reducing disease incidence in pumpkin plants.

5 白粉病菌对结果期南瓜叶片光合特性和叶绿体超微结构的影响

病原菌侵染时，植物光合效率降低，通常与光合器官受损、光合代谢所需激发能过量增加有关，最终导致氧气的浓度下降和活性氧（reactive oxygen species，ROS）产生（付健等，2021；许珂等，2021；Iqbal et al.，2012）。过量ROS积累间接诱导植物的过敏性反应（hypersensitive reaction，HR）（Petrov et al.，2012）。HR是植物与病原菌不亲和互作发生的一种抗病反应，表现为在病原菌侵染处形成不同于周围健康组织的局部细胞坏死斑（向婧姝，2018），是细胞程序性死亡的主要形式，可以激发邻近组织的防卫反应和植株的系统获得性抗性（systemic acquired resistance，SAR）（徐正，2018）。甜瓜幼苗感染白粉病后光合指标下降，包括净光合速率（P_n）、气孔导度（G_s）、胞间CO_2浓度（C_i）、蒸腾速率（Tr）（刁倩楠等，2021）。张兆辉等（2021）研究发现，接种白粉病菌的西葫芦叶片叶绿体结构发生类囊体膨大、基质片层和基粒片层扭曲等变化。白粉病降低植物的光合作用，但关于白粉病菌影响南瓜光合特性的机制报道很少。鉴于南瓜果实发育期开花和幼果生长阶段易暴发白粉病，以白粉病不同抗性南瓜品种为试材，研究白粉病菌对南瓜结果期叶片光合指标、细胞坏死及叶绿体超微结构的影响，分析其光合特性变化的生理机制，以期丰富南瓜抗白粉病的防御机制研究。

5.1 材料和方法

5.1.1 供试材料

供试南瓜材料为中国南瓜品种'九江轿顶'（JJJD）和自交系112-2，其中'九江轿顶'高感白粉病，自交系112-2高抗白粉病。2001年至今，在实验室人工接种白粉病菌，自交系112-2幼苗期植株均表现出稳定的高抗白粉病特性（Chen et al.，2020；Zhou et al.,2010)。在田间自然发病条件下,南瓜结果期自交系112-2整株基本未出现白粉病症状,而相邻定植的'九江轿顶'中下部叶片布满大小不一的白粉斑,发病严重的基部叶片出现萎黄、早衰症状,并向上蔓延。自交系112-2和'九江轿顶'种子均由河南科技学院南瓜课题组提供。

5.1.2 试验设计

2020年4月在河南科技学院育种实验室将供试南瓜种子浸种催芽后播种于穴盘内,幼苗长至2叶1心时移栽到口径为30 cm的大盆中,将大盆搬至室外进行盆栽管理,植株进入结果期,即第1朵雌花开花后,进行白粉病菌接菌处理。采集感白粉病的南瓜叶片,用软毛刷将白粉病菌孢子扫入培养皿内,无菌水稀释,调整孢子悬浮液终浓度为$1×10^5$个/mL。采用喷雾法将白粉病菌孢子悬浮液喷洒于结果期南瓜第1朵雌花上下的2片叶片,以叶片布满雾滴状水珠但不滴落为宜,CK喷清水(不接种白粉病菌)。以112-2-PM、JJJD-PM分别表示自交系112-2、'九江轿顶'接种白粉病菌处理,112-2-CK、JJJD-CK分别表示自交系112-2、'九江轿顶'清水对照处理。

5.1.3 光合参数测定

在自然光照条件下,随机选取清水(CK)和白粉病菌处理的结果期南瓜植株各9株,每株选择雌花附近、长势一致的2片叶片做标记,于上午9:00—11:00用Li-6400便携式光合仪测定南瓜叶片的P_n、G_s、T_r和C_i,从接种白粉病菌当天开始,间隔1 d测1次,至接种后第12天结束,即0、2、4、6、8、10、12 d。每个叶片重复测定3次。

5.1.4 坏死斑观察

分别在接种白粉病菌后0、4、5、6 d随机选取2种南瓜材料各12株,每株选择雌花附近、长势一致的叶片,避开叶脉,剪取南瓜叶片,样本大小1 cm×1 cm,设置3个重复。将剪取的南瓜叶片浸没在台盼蓝染色液中,沸水浴2 min后,乙醇脱色至透明。将脱色后的叶片制成玻片,在倒置荧光显微镜下观察,细胞坏死斑被台盼蓝染色剂染成蓝色。每个样品取 10 个视野。

5.1.5 叶绿素含量测定

参照蔡庆生（2013）的方法。分别于接种白粉病菌后 0、4、8、12 d 随机选取 2 种结果期南瓜植株各 12 株，每株选择雌花附近、长势一致的叶片，避开叶脉剪碎混匀，称取 0.2 g 放于研钵中，倒入 95% 乙醇及少量碳酸钙，研磨成匀浆，静置 3～5 min 后过滤，95% 乙醇定容至 25 mL，摇匀，设置 3 次重复。计算叶绿素 a（C_a）含量、叶绿素 b（C_b）含量以及总叶绿素（C_T）含量。

$$C_a \text{ (mg/g)} = 13.95\ A_{665} - 6.88\ A_{649},$$
$$C_b \text{ (mg/g)} = 24.96\ A_{649} - 7.32\ A_{665},$$
$$C_T \text{ (mg/g)} = 6.63\ A_{665} + 18.08\ A_{649}。$$

5.1.6 叶绿体超微结构观察

接种白粉病菌后 4 d，选取自交系 112-2 和'九江轿顶'南瓜各 3 株，取上述相同叶位的叶片，以健康无病害的叶片作 CK。用锋利徕卡刀片避开主叶脉切取 0.5 mm×1.0 mm 的小块，快速投入装有预冷戊二醛（4%）溶液的青霉素小瓶中，用注射器抽气至材料大部分下沉于瓶底部，4℃下固定 6 h；用磷酸缓冲液（0.1 mol/L）冲洗 3 次，每次 30 min；四氧化锇固定液二次固定 6 h，磷酸缓冲液再次冲洗；用不同体积分数的（15%、30%、50%、70%、80%、90%、100%）乙醇连续脱水 2 次，每次 30 min；丙酮过渡，环氧树脂渗透包埋，在 35℃ 12 h、45℃ 12 h、60℃ 24 h 条件下分别聚合；用 LEICA EM UC7 型超薄切片机钻石刀切片，切片厚 60～80 nm，25℃下用醋酸双氧铀-柠檬酸铅进行双重染色。于 HITACHI-HT7700 型透射电子显微镜下观察拍照，每个样品取 10 个视野。

5.1.7 数据处理与分析

所有数据均取 3 次重复的平均值，数据分析和作图分别用 SPSS 2.0 软件和 Excel 2003 进行。

5.2 结果与分析

5.2.1 白粉病菌对南瓜结果期叶片光合指标的影响

图1-9显示,接种白粉病菌后,2种材料(自交系112-2和'九江轿顶')叶片的P_n、T_r和G_s均低于清水对照,'九江轿顶'的P_n下降幅度总体上大于112-2,其中,自交系112-2和'九江轿顶'叶片P_n在接种后2 d分别比相应对照降低45.4%和63.8%;但接种后6~10 d,2种材料叶片的C_i则高于对照。抗性材料112-2接菌后总体上表现为P_n高于感病材料'九江轿顶'接菌处理,C_i低于感病材料'九江轿顶'接菌处理。可见,白粉病会影响结果期南瓜叶片的光合效率,降低P_n、T_r和G_s,增加C_i,说明白粉病菌侵染降低了叶片对C_i的利用率,此外,感病材料中这种影响高于抗病材料。

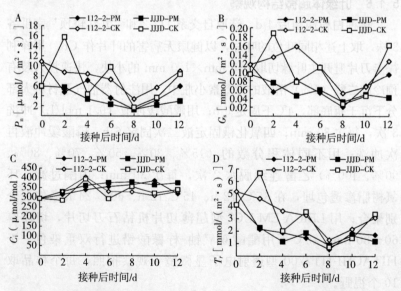

图1-9 白粉病菌对南瓜材料自交系112-2和'九江轿顶'结果期叶片光合指标的影响

5.2.2 白粉病菌对南瓜结果期叶片叶绿素含量的影响

图 1-10 显示，与 CK 相比，接种白粉病菌后，'九江轿顶'的 C_a、C_b 和 C_T 含量在接菌前期（接种后 0~8 d）无明显变化，而在接种后第 12 天明显降低，与 CK 相比分别降低 50.0%、47.0% 和 49.0%；材料 112-2 C_a、C_b 和 C_T 含量在接种后 4 d 和 12 d 明显低于 CK，其中接种后 12 d 分别下降 65.0%、54.0% 和 57.0%，降幅明显高于感病材料'九江轿顶'。表明白粉病菌侵染降低了南瓜结果期叶片 C_a、C_b 和 C_T 的含量，其中 C_a 含量的下降最大，且抗病材料叶绿素含量的下降幅度较大。

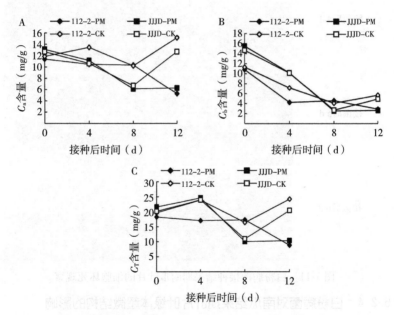

图 1-10　白粉病菌对南瓜材料自交系 112-2 和'九江轿顶'结果期叶片叶绿素含量的影响

5.2.3 白粉病菌对结果期南瓜叶片细胞坏死的影响

图 1-11 显示，接种白粉病菌后 4 d，自交系 112-2 和'九江轿顶'均出现坏死斑，随着接种后时间的延长，坏死斑面积增大、颜色加深，感病材料'九江轿顶'坏死斑面积明显大于抗病材料自交

系112-2且颜色较深，表明接种白粉病菌诱发结果期南瓜叶片出现HR-细胞坏死，感病材料HR发生较严重。

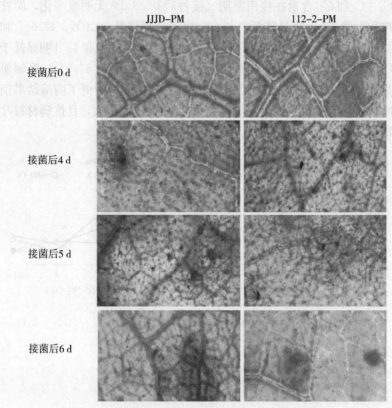

图1-11 白粉病菌接种结果期南瓜叶片的细胞坏死观察

5.2.4 白粉病菌对南瓜结果期叶片叶绿体超微结构的影响

图1-12显示，清水处理南瓜材料（自交系112-2和'九江轿顶'）叶绿体呈现长椭圆形，紧靠细胞膜分布，叶绿体膜结构清晰平滑，基质片层和基粒片层排列紧密，片层垛叠整齐均匀，有少量电子密度高的嗜锇体存在。接种白粉病菌后4 d，自交系112-2叶片叶绿体大多肿胀变形，并向细胞中央游离，基质片层排列疏松，基粒片层不规则排列，嗜锇体增多，积累较多淀粉粒；'九江轿顶'

叶片叶绿体的肿胀程度大于自交系112-2，叶绿体间挤压严重，细胞膜、基质片层、基粒片层模糊不清，淀粉粒较多，并形成大量油滴。

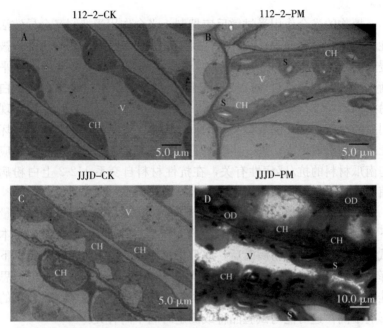

图 1-12　结果期南瓜接种白粉病菌后叶绿体超微结构的观察
CH. 叶绿体；S. 淀粉粒；OG. 嗜锇体；OD. 油滴

5.3　结论与讨论

植物叶片光合系统对白粉病非常敏感。金海军等（2020）研究表明，在感染白粉病菌的 17 份黄瓜材料中大多数材料 P_n、Tr、G_s 明显降低，C_i 增加。本研究也表明，结果期接种白粉病菌的南瓜叶片 P_n、Tr、G_s 均降低，而 C_i 增加。通常，气孔关闭和叶肉细胞光合活性下降导致的非气孔因素是病原菌胁迫引起植物光合作用下降的主要影响因素。Farquhar 等（1982）研究表明，G_s 下降，C_i 不变或上升，光合速率下降是由叶肉细胞同化能力降低等非气孔因素引起的。本试验结果显示，结果期接种白粉病菌后，南瓜叶片的 G_s 下降，C_i 上升，说明非气孔因素是南瓜叶片光合作用降低的主要原因之一，这与李宁等（2013）和王莹等（2020）的研究结果一致。

植物的光合作用受过敏反应抑制，光合作用受抑制又会导致活性氧产生（秦宏坤等，2020），活性氧影响叶绿体 PSⅡ 的活性，被侵染的细胞由于 PSⅡ 活性丧失，干扰了细胞存活所必需的动态平衡，反过来加速了过敏反应（沈喜等，2003）。本试验中，接种白粉病菌后 4 d，南瓜（自交系 112-2 和'九江轿顶'）叶片均出现明显的过敏反应，随着接种后时间的延长，感病材料 JJJD 细胞坏死斑面积扩展更大，进一步限制了叶片的光合速率，说明过敏反应与南瓜材料的抗/感病性有关，在抗性材料自交系 112-2 上白粉病菌的生长受到有效抑制（张兆辉等，2021）。

病原菌侵染植物后可破坏叶绿体的结构和功能，使叶绿素含量下降。甜瓜感染白粉病后叶绿素含量下降（李小玲等，2015）。本试验发现，接种白粉病菌后，南瓜材料 C_a、C_b 和 C_T 含量均明显下降，且抗性材料自交系 112-2 叶绿素含量的下降幅度较大，与柯思佳等（2019）研究认为抗性甜瓜材料能减缓蔓枯病对叶绿素含量的影响不同，这可能与试验品种或测定时间不同有关。

叶绿体是植物光合反应的主要场所，其结构的稳定直接影响光合作用的强弱（李冬林等，2019），甘薯感染疮痂病后叶绿素被降解，叶绿体超微结构发生变化，叶绿体膜流动性降低，感病品种叶绿体受损程度远大于抗病品种（Yu et al.，2006）。本试验发现，南瓜感染白粉病菌后叶绿体肿胀变形，基质片层排列疏松，基粒片层不规则排列，嗜锇体和淀粉粒增多，且感病材料叶绿体结构受损更为严重。病原菌侵染导致植株产生大量 ROS，引起叶绿体膜过氧化（何金环等，2005），造成叶绿素降解，致使植物光合能力下降。ROS 对膜系统的损伤进一步阻碍了淀粉的水解和向外运输，导致叶绿体中大量淀粉粒累积，淀粉的这种积累称为被动存储（赵海新等，2020）。大量淀粉颗粒的产生挤压叶绿体，使叶绿体肿胀加剧，南瓜接种白粉病菌后叶绿体中嗜锇体数量增多，表明叶绿体膜受到了过氧化损伤。感病材料'九江轿顶'叶绿体中出现大量油滴，表明叶绿体膜质过氧化程度加深，对叶绿体结构造成了不可逆的损伤。总之，结果期白粉病菌侵染引起南瓜叶片光合性能下降，主要归因于侵染点细胞坏死、叶绿素含量下降和叶绿体结构受到破坏，抗病材料自交系112-2光合速率降幅较小与细胞坏死和叶绿体结构受损程度较轻有关。

第2章 南瓜抗白粉病相关基因的分离及功能研究

1 Improved powdery mildew resistance of transgenic tobacco overexpressing the *Cucurbita moschata CmSGT1* gene

1.1 Introduction

The genus *Cucurbita* is composed of several species, including the cultivated *C. moschata* (*Cucurbita moschata* Duch.), *C. pepo* (*Cucurbita pepo* L.), *C. maxima* (*Cucurbita maxima* Duch.) and several wild species. Pumpkins (*C. moschata*) are valued for their fruit and seeds. Additionally, they are rich in vitamins, amino acids, flavonoids, phenolics, and carbohydrates, and possess medicinal properties, including anti-diabetic, anti-oxidant, anti-carcinogenic, and anti-inflammatory activities (Wang et al., 2002; Yadav et al., 2010). In China alone, the annual yield of pumpkin, squash and gourds is 8 051 495 t (i.e., approximately 22.68% of the global yield) from a harvested area of 438 466 ha (i.e., 17.42% of the global area) (FAO, 2017). Cucurbit powdery mildew (PM) is a serious disease affecting field-grown and greenhouse-grown cucurbit crops worldwide. The disease is mainly caused by *Podosphaera xanthii* (formerly known as *Sphaerotheca fuliginea*), which is a biotrophic plant pathogen (Fukino et al., 2013; Perez-Garcia et al., 2009). Fungicide applications poorly control PM and the long-term use of pesticides may lead to increased environmental pollution and the residual chemicals on food crops may be harmful for humans and animals. Therefore, studying the mechanism underlying PM resistance and exploiting the resistance genes to breed resistant varieties represents an effective way to control PM in pumpkin.

To date, there has been relatively little research on pumpkin ($2n = 2x = 40$), especially at the molecular level, which has seriously hindered developments in the fields of molecular biology and genetics. We previously conducted a RNA sequencing analysis of pumpkin inbred line highly resistant to PM (inbred line112-2) and identified 4 716 differentially expressed genes, including genes encoding broad-spectrum disease-resistance/susceptibility proteins [PR protein, ubiquitin, heat shock protein and Mildew Locus O (MLO)], reactive oxygen scavengers (peroxidase, superoxide dismutase, and catalase), signal transduction molecules serine/threonine protein kinase, and transcription factors (WRKY, TGA, and MYB). Additionally, candidate disease-resistance genes, such as *WRKY21*, *MLO3*, and *SGT1*, were identified (Guo et al., 2018). One of the genes cloned from the transcriptome was a homolog of *Cucumis melo SGT1* (*suppressor of the G2 allele of skp1*). The expression of this *SGT1* homolog was highly upregulated in PM-resistant material at 6 h after an inoculation with the PM fungus, but not in PM-susceptible material. However, pumpkin *SGT1* has not been functionally characterized regarding its potential involvement in the defense response to PM.

The SGT1 protein was originally defined in yeast, in which it interacts with SKP1, which is a component of the Skp1/CDC/F-box protein E3 ubiquitin ligase complex (Kitagawa et al., 1999). The SGT1 protein contains three functional domains, namely the tetratricopeptide repeat (TPR), CHORD SGT1 (CS), and the SGT1-specific sequence (SGS) (Kumar and Kirti, 2015). SGT1 is closely related to the disease resistance mediated by the plant resistance (*R*) genes. The silencing or mutation of *NbSGT1* from *Nicotiana benthamiana* can lead to reduction of steady-state levels of R proteins and the loss of the mediated resistance, and *AtSGT1a* overexpression can contribute positively to resistance triggered by the NB-LRR type R proteins, and can complement for loss of *AtSGT1b* in auxin signalling (Muskett and Parker, 2003; Azevedo et al., 2006). A previous study revealed that *NbSGT1* overexpression in *N. benthamiana* accelerates the development of the hypersensitive response (HR) during *R*-mediated disease resistance (Wang et al., 2010). Additionally, *Hv-SGT1* overexpression in wheat enhances the resistance to PM, which is correlated with increased levels of reactive oxygen intermediates at the pathogen entry sites (Xing et al., 2013). Furthermore, *PsoSGT1* from *Prunus sogdiana* appears to interact with molecular chaperones (RAR1 and HSP90) to activate a nucleotide-binding domain and leucine-rich repeat-containing (NB-LRR) -type protein that confers disease resistance (Zhu et al., 2017). The regulation of NB-LRR-type protein stability or substrate degradation to maintain the balance between the activation and inhibition of plant defense responses (Hoser et al., 2013; Meldau et al., 2011), contributes to the RXLR elicitor-induced HR-related cell necrosis (Xiang et al., 2017).

The recent release of the pumpkin genome provided an opportunity for an additional screening for disease-resistance genes (Sun et al., 2017). In this study, on the basis of the above-mentioned transcriptome analysis that identified differentially expressed genes responsive to PM, we functionally characterized the pumpkin homolog of *SGT1* (designated as *CmSGT1*). The transcription of *CmSGT1* in the PM-resistant inbred line 112-2 was strongly induced by PM, salicylic acid (SA), hydrogen peroxide (H_2O_2), ethephon (Eth), and methyl jasmonate (MeJA). Transgenic tobacco plants that constitutively overexpressed *CmSGT1* exhibited increased resistance to PM and were hypersensitive to bacterial wilt and scab.

1.2 Results

1.2.1 Cloning of *CmSGT1* and subcellular localization the encoded protein

Pumpkin PM-related candidate genes identified in a transcriptome were reported previously (Guo et al., 2018). One of the isolated clones exhibited 89% identity at the nucleotide level to *C. melo SGT1*. A full-length clone of this homolog was obtained, and the gene was named *CmSGT1* and submitted to GenBank (accession number MH105820). The *CmSGT1* gene comprised 1 206 bp, which included a 1 080 bp ORF encoding 360 amino acids. The predicted polypeptide was basic, with a pI of 5.32, and a molecular mass of 40.3 ku. An alignment of the deduced CmSGT1 amino acid sequence with homologous sequences is presented in Fig. 2-1. At the amino acid level, *CmSGT1* was highly similar to the *SGT1* from the following plant species: *Cucumis melo* (*CmSGT1*, 84.4% identity), and *Cucumis sativus* (*CsSGT1*, 82.5% identity), *N. benthamiana* (*NbSGT1.1* and *NbSGT1.2*, 89.5% identity), and *A. thaliana* (*AtSGT1a* and *AtSGT1b*, 75.2% identity). The *CmSGT1* sequence contained three conserved domains, namely TPR, CS, and SGS.

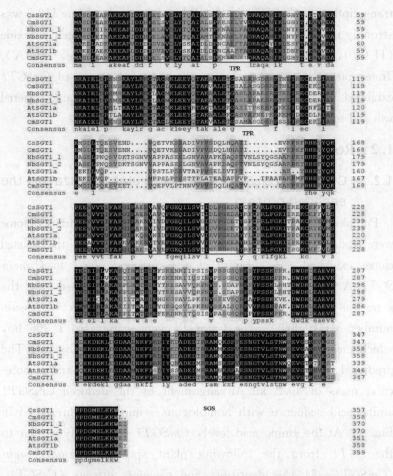

Fig. 2-1 Amino acid sequences alignment of pumpkin CmSGT1 with others. The three conserved domains (TPR, CS, and SGS) are shown by thin underlines. The genes included are CmSGT1 (*Cucumis melo* L., XP_008439299.1) and CsSGT1 (*Cucumis sativus* L., XP_004140745.1) AtSGT1A (Arabidopsis, AT4G23570.3) and AtSGT1B (Arabidopsis, AT4G11260.1), NbSGT1.1 (*N. benthamiana*, AF516180) and NbSGT1.2 (*N. benthamiana*, AF516181).

The subcellular localization of *CmSGT1* was assessed with a CmSGT1-GFP fusion protein that was produced in *A. thaliana* protoplasts under the control of the 35S CaMV promoter. The GFP signal in protoplasts producing GFP alone was detected in the cytoplasm and nucleus (Fig. 2-2), whereas the signal from the CmSGT1-GFP fusion protein was detected exclusively in the nucleus.

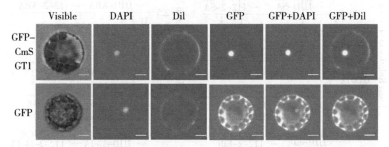

Fig. 2-2   The subcellular localization of pumpkin *CmSGT1*. The fused pBI221-GFP-CmSGT1 and pBI221-GFP constructs were introduced into Arabidopsis protoplast by polyethylene glycol (PEG) -mediated protoplast transformation. The fluorescent signals were detected using a confocal fluorescence microscope. Scale bars=5 μm.

1.2.2 Expression of *CmSGT1* in pumpkin seedlings in response to PM and exogenous treatments

A qRT-PCR assay was completed to analyze the *CmSGT1* expression patterns in PM-resistant inbred line 112-2 and PM-susceptible cultivar "JJJD" treated with PM, H_2O_2, SA, ABA, Eth, and MeJA (Fig. 2-3). The expression data were normalized against that of the *β-actin* gene and were recorded relative to the *CmSGT1* transcript level in the water-sprayed control "JJJD" plants at 0 hour post-inoculation (hpi).

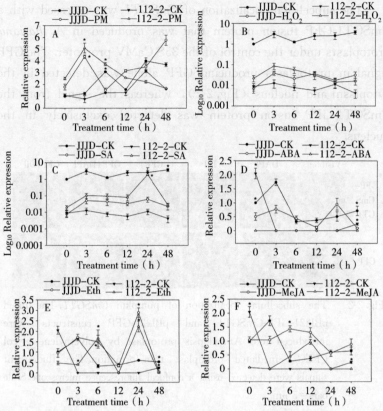

Fig. 2-3 *CmSGT1* expression in response to powdery mildew and exogenous hormones. The pumpkin seedlings were sprayed with a spore suspension (A), exogenous H_2O_2 (B), SA (C), ABA (D), Eth (E) and MeJA (F). The pumpkin *β-actin* gene was used as an internal reference gene for qRT-PCR. The transcript level of *CmSGT1* in the susceptible cultivar JJJD at 0 h is used as control (quantities of calibrator) and was assumed as 1. The relative gene expression in B and C (Y-axis) was transformed to a $\log_{10}$ scale. The values are the means ± SEs of three biological replicates. Data between treatments (112-2-treatment vs 112-2-CK and JJJD-treatment vs JJJD-CK) was analyzed by one-way ANOVA, and * denotes statistical significance at $P < 0.05$.

The *CmSGT1* expression level in 112-2 plants was upregulated by PM (except at 24 h), H_2O_2, and SA treatments, with the PM-induced expression level at 3 hpi upregulated by 6.62-fold. In contrast, the ABA treatment essentially had no effect. The expression of *CmSGT1* in "JJJD" plants was inhibited by H_2O_2, SA, and ABA (except at 24 h) treatments, with irregular expression-level changes induced by PM. After the Eth treatment, the *CmSGT1* expression level was significantly higher in the 112-2 and "JJJD" seedlings than in the water-treatment control (CK) seedlings (except at 48 h), with peak levels occurring at 24 h (i.e., 10.2-fold and 4.5-fold increases in 112-2 and "JJJD" seedlings, respectively). In response to the MeJA treatment, *CmSGT1* expression levels were higher in 112-2 (except at 48 h) and "JJJD" (except at 3 h) seedlings than in CK seedlings. The results indicated that SA, and H_2O_2 upregulated *CmSGT1* expression in inbred line 112-2 (PM-resistant material), but downregulated *CmSGT1* expression in cultivar "JJJD" (PM-sensitive material). Moreover, Eth and MeJA induced *CmSGT1* expression regardless of PM susceptibility.

1.2.3 Improved PM resistance of *CmSGT1*-overexpressing tobacco plants

Firstly, transcript level of the high homology (*NbSGT1*) modulated by the overexpression of *CmSGT1* under normal conditions was determined by qRT-PCR. Compared to wild type (WT) plants, the expression of the homologous genes was not basically altered in transgenic plants when grown in normal condition (Fig. 2-4), indicating that overexpression of pumpkin *CmSGT1* gene has no obvious effect on *NbSGT1* transcript in tobacco.

The *CmSGT1* gene was not detected in WT plants.

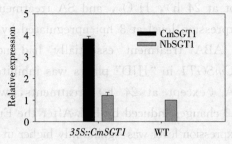

Fig. 2-4 Relative expression of *CmSGT1* and *NbSGT1* in transgenic or wild-type plants under normal growth conditions. *N. benthamiana* encodes *NbSGT1* isoform. L3, L5, and L11 are three independent transgenic lines that overexpress *CmSGT1*. Data (*NbSGT1* expression in transgenic lines vs *NbSGT1* expression in wild-type plants) were analyzed by one-way ANOVA, and unmarked * denotes no statistical differences at $P < 0.05$.

The disease severity at 10 days post inoculation (dpi) was 78% lower for the transgenic plants than for the CK plants (Table 2-1). Powdery mildew symptoms were detectable on infected WT seedlings at 7 dpi, and the infected spots were chlorotic at 28 dpi. In contrast, the PM symptoms were undetectable and relatively slight at 7 and 28 dpi, respectively, in the *CmSGT1*-overexpressing transgenic plants (Fig. 2-5A). A comparison of the leaves of PM-infected WT and transgenic plants revealed more blue spots displayed by the trypan blue staining, on the transgenic leaves than on the WT leaves at 4 dpi, although blue spots were developing on the WT leaves. The blue spots on the transgenic leaves expanded at 5 and 7 dpi, and were bigger than those of WT leaves, implying that the overexpression of *CmSGT1* in transgenic plants accelerated cell death following a PM infection (Fig. 2-5B). Moreover, there were more brown spots manifested by the DAB staining, on the transgenic leaves than on the WT leaves at 1 dpi. These brown spots reflected the accumulation of H_2O_2, and were darker

第 2 章 南瓜抗白粉病相关基因的分离及功能研究

brown and larger at 3 dpi. At 5 dpi, the brown spots lightened in the infected plants, but were more intense in the transgenic plants than the spots in the WT plants. The results indicated that the overexpression of *CmSGT1* promoted the accumulation of H_2O_2 in transgenic plants infected with PM (Fig. 2-5C).

Table 2-1 Disease severity of leaves of tobacco seedlings infected with powdery mildew

Materials	Disease severity (in vitro leaf)
L3	7.90±1.23
L5	8.60±1.31
L11	9.00± 1.01
WT	38.80± 2.41*

Note: Data are mean values (±SD) of at least three independent experiments. * indicates significantly different values between treatments ($P<0.05$). L3, L5, and L11 are three independent transgenic lines that overexpress *CmSGT1*.

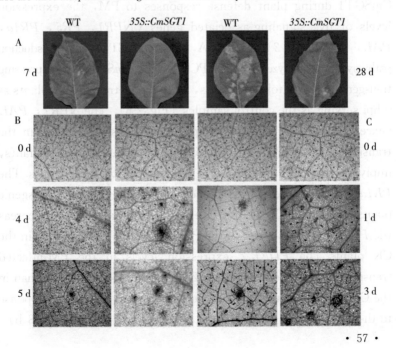

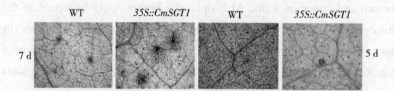

Fig. 2-5　The pathogenic symptoms, trypan blue and DAB staining of tobacco leaves treated with powdery mildew. The pathogenic symptoms of transgenic tobacco and WT at 7, and 28 day post inoculation (A); trypan blue staining was performed using the *Agro*-infiltration method to visualize cell death (B); DAB staining was performed using the *Agro*-infiltration method to visualize H_2O_2 accumulation (C). Scale bars=200 μm.

1.2.4　Expression of signal-related genes in transgenic tobacco plants

To investigate the signal transduction pathways affected by CmSGT1 during plant defense responses to PM, the expression levels of five signaling-associated genes (*NPR1*, *PR5*, *PR1a*, *PAL*, and *PDF1.2*) in the SA, JA, and ET signal transduction pathways were analyzed by qRT-PCR for the *CmSGT1*-overexpressing transgenic and WT tobacco plants, with water-treated WT plants at 0 hpi serving as the control samples (Fig. 2-6). The *NPR1*, *PAL* (except at 120 h), and *PR5* expression levels were lower in the transgenic and WT plants infected with PM than in the CK plants, implying that PM inhibited the expression of these genes. The *PR1a* expression level was higher in the PM-infected transgenic plants than in the CK plants (i.e., 68.3-fold at 120 h), whereas the *PR1a* level was lower in the PM-infected WT plants than in the CK plants. The *PDF1.2* expression levels in the PM-infected transgenic plants were lower at 12 hpi and higher at 48 hpi than in the CK plants, with no differences thereafter. Thus, in response to the PM infection, the *NPR1*, *PAL* (except at 24 and 48 h),

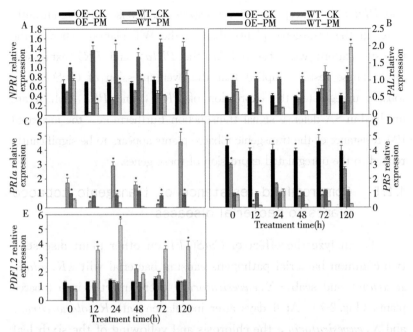

Fig. 2-6　Expression of signal-related genes in transgenic and wild-type plants treated with powdery mildew. Relative expression levels of signal-related genes were determined by qRT-PCR using cDNA synthesized from total RNAs isolated from the mixed sample of three genetically modified tobacco lines treated with powdery mildew at different time points. A, NtNPR1; B, NtPAL; C, NtPR1a; D, NtPR5; E, NtPDF1.2. There were four treatments: OE-CK represents transgenic plants grown under normal conditions; OE-PM represents transgenic plants infected with powdery mildew; WT-CK represents wild-type plants grown under normal conditions; WT-PM represents wild-type plants infected with powdery mildew. Tobacco $NtEF1\text{-}\alpha$ gene (AF120093) was used as an internal control for normalization of different cDNA samples. The expression levels of signal-related genes in wild-type plants at 0 h were used as control (quantities of calibrator) and were assumed as 1. Three biological triplicates were averaged and Bars indicate standard error of the mean. Data between transgenic plants and WT plants (OE-PM vs WT-PM and OE-CK vs WT-CK) were analyzed by one-way ANOVA, and * denotes statistical significance at $P<0.05$.

and *PDF1.2* (except at 48 h) expression levels were significantly lower in the transgenic plants than in the WT plants, whereas the opposite pattern was observed for the *PR1a*, and *PR5* expression levels. These results implied that the overexpression of *CmSGT1* in tobacco upregulates the expression of *PR1a*, and *PR5*, which are involved in SA defense signal transduction. Furthermore, the increased PM resistance of the transgenic tobacco plants appears to be significantly related to the upregulated expression of these genes.

1.2.5 Compromised resistance of transgenic tobacco plants to bacterial diseases

To analyze the effect of *CmSGT1* on other plant diseases, two common bacterial pathogens causing bacterial wilt (*R. solanacearum*) and scab (*X. euvesicatoria*) were injected into tobacco plants (Fig. 2-7). At 6 days after inoculation with *R. solanacearum* and *X. euvesicatoria*, the chlorosis and yellowing of the sixth leaf veins was greater for transgenic tobacco plants than for the WT

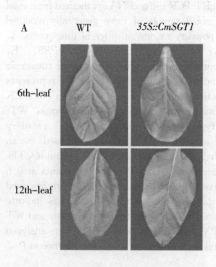
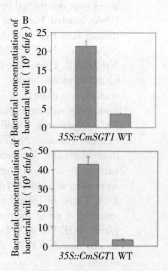

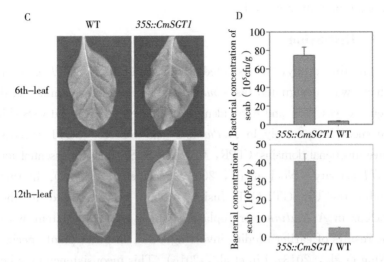

Fig. 2-7　The resistance of *CmSGT1* in transgenic and wild-type plants to tobacco bacterial wilt and scab. A, pathogens symptoms of the 6th-upper and 12th-upper leaf injection sites was injected with bacterial wilt bacteria with a needle-removed syringe; B, concentration bacteria of the 6th-upper and 12th-upper leaf injection sites was injected with bacterial wilt bacteria with a needle-removed syringe; C, pathogens symptoms of the 6th-upper and 12th-upper leaf injection sites was injected with scab bacteria with a needle-removed syringe; D, concentration bacteria of the 6th-upper and 12th-upper leaf injection sites was injected with scab bacteria with a needle-removed syringe. Significant differences between WT and transgenic plants at $P<0.01$.

plants. Additionally, there were 5.94 and 21.1 times more the concentration of bacterial wilt and scrab bacteria, respectively, in the transgenic plants than in the WT plants. Yellowing was also observed between the veins of the 12th leaf in transgenic tobacco plants, and there were 13.3 and 8.28 times more *R. solanacearum* and *X. euvesicatoria* bacteria, respectively, in the transgenic plants than in the WT plants. These observations suggested that the overexpression of *CmSGT1* in tobacco decreases the resistance to bacterial wilt and scrab.

1.3 Discussion

In this study, we isolated a novel pumpkin *SGT1* gene, which was designated as *CmSGT1*. The predicted amino acid sequence was 89% and 82% identical to the NbSGT1.1 and CsSGT1 sequences, respectively. In *A. thaliana* and tobacco, SGT1 encodes three functional domains (TPR, CS, and SGS) that are essential for *SGT1* activity (Noël et al., 2007; Peart et al., 2002). In this study, the CmSGT1-GFP fusion protein was localized to the nucleus in *A. thaliana* protoplasts, which was inconsistent with the results of earlier studies involving SGT1 in other plant species (Xing et al., 2013; Liu et al., 2016). This inconsistency may be related with low identity (*Haynaldia villosa* HvSGT1, 64.0% identity; pepper CaSGT1, 63.7% identity) or the number of transformed cells examined. *Arabidopsis SGT1b* fused to Cerulean localized to the cytosol, but could be seen in nuclei of 25% of 55 transformed cells examined (Noël et al., 2007), and translocation of the SGT1/SRC2-1 complex from the plasma membrane and cytoplasm to the nuclei is required in pepper upon the inoculation of *P. capsici* (Liu et al., 2016), suggesting the movement of *SGT1* between the cytosol and nucleus.

The interplay among complex signaling networks, including various pathways regulated by phytohormones, such as SA, JA, ethylene (ETH) and ABA, considerably influences plant resistance to diseases. An earlier investigation proved that *HvSGT1* expression levels substantially increase following treatments with H_2O_2 and MeJA, slightly increase following exposure to ETH or ABA, and are unchanged in response to SA (Xing et al., 2013). In the current study, SA and H_2O_2 treatments considerably upregulated *CmSGT1* expression in the PM-resistant inbred line 112-2, but downregulated *CmSGT1* expression in the PM-susceptible material. The expression of *CmSGT1* in both materials was upregulated by ETH and MeJA treatments. Recent reports indicated that *SGT1* expression may be induced by phytopathogens, such as *Blumeria graminis* in wheat and *Phytophthora capsici* in pepper (Xing et al., 2013; Liu et al., 2016). In this study, the *CmSGT1* expression in inbred line 112-2 was induced by PM. Additionally, the overexpression of *CmSGT1* in transgenic tobacco plants decreased the disease index by 78%, accelerated cell necrosis, and increased the accumulation of H_2O_2. These results indicated that the PM resistance of the transgenic tobacco plants was enhanced, likely because of the changes to the HR-related cell necrosis and H_2O_2 accumulation. Our findings are consistent with the results of an earlier study that revealed that H_2O_2 accumulation and the subsequent cell death usually lead to the resistance to diseases caused by biotrophic pathogens (Li et al., 2011). Moreover, *SGT1* helps mediate cell death during compatible and incompatible plant-pathogen interactions, suggesting that *SGT1* is an essential component of common signaling pathways responsible for cell death (Uppalapati et al., 2010; Cuzick et al., 2009; Wang et al., 2010). The silencing of the pepper *SGT1* gene adversely affects HR-related cell death, prevents H_2O_2 accumulation, and downregulates HR-related and SA/JA-dependent marker gene expression levels, and influences the PcINF1/SRC2-1-induced pepper defense response by *SGT1* interacting with SRC2-1 (Liu et al., 2016).

The *PDF1.2* gene is important for the JA/ETH-dependent signaling pathway. The disease-associated *NPR1* gene (non-expresser of PR1) affects various disease-resistance signal transduction pathways, and encodes one of the important transcription factors downstream of SA. The SA-dependent disease-resistance signal transduction pathway can be divided into NPR1-dependent and NPR1-independent transduction pathways (Gao and Zhang, 2012). The *PAL*, *PR1a*, and *PR5* expression levels are markers of the SA signaling pathway. In wheat, HvSGT1 activates PM resistance mechanisms through JA-dependent defense pathways and suppresses the activities of SA-dependent defense pathways (Xing et al., 2013). In the current study, following a PM infection, the *NPR1*, *PAL*, and *PDF1.2* expression levels were lower in the transgenic plants than in the WT plants, whereas the opposite pattern was observed for the *PR1a*, and *PR5* expression levels. This suggests that in the SA pathway, the transactivation of *NPR1* is unaffected by CmSGT1, whereas the transactivation of *PR1a*, and *PR5* is dependent on *CmSGT1*. Additionally, CmSGT1 does not directly affect the JA/ETH-dependent defense pathway to regulate *PDF1.2* expression. We propose that *CmSGT1* activates stress-resistance mechanisms through SA-dependent defense pathways without inducing *NPR1*, and *PAL* expression levels, but suppresses the activities of JA/ETH-dependent defense pathways. Therefore, we speculate that CmSGT1 positively regulates the H_2O_2 and SA pathways. Moreover, H_2O_2 might directly transfer the SA signal to regulate the expression of downstream response genes in the *CmSGT1*-overexpressing transgenic plants infected with PM. As a NPR1-dependent SA signal transduction pathway, SA-mediated activation of *PR* genes is required for binding activity of NPR1 and TGA transcription factors to promoter elements (Spoel and Dong, 2012; Fan and Dong, 2002). NPR1 mutant losses the function in binding SA, it promotes SA-induced defense gene expression (Ding et al., 2018). We speculated that whether there is the relation of the downregulation of *NPR1* and SA-mediated transcriptional activation of *PR* genes. Notablely, the phenotypes and genes overexpressed by *CmSGT1* in transgenic tobacco plants challenged with tobacco PM might not exactly the same as those regulated by *CmSGT1* in *C. moschata* in response to cucurbit PM. Further studies will be necessary to reveal biological functions for *CmSGT1* in *C. moschata* infected with PM.

Two globally important diseases that affect tobacco, bacterial wilt and scab, are caused by the necrotrophic *Ralstonia solanacearum* and the hemi-biotrophic *Xanthomonas euvesicatoria* respectively. There are differences in the resistance mechanisms and patterns of fungal development for biothophic (*P. xanthii*) and necrotrophic (*R. solanacearum*) pathogens. Recent studies confirmed that SGT1 promotes the resistance to biotrophic pathogens, while suppressing the resistance to necrotrophic and hemibiotrophic pathogens. Silencing of *SGT1* compromised resistance to the barley and *Haynaldia villosa* biotroph *Blumeria graminis* (Shen et al., 2003; Xing et al., 2013) and the wheat biotroph *Puccinia striiformis* (Scofield. et al., 2005), while enhancing the resistance of tobacco to the necrotrophic pathogen *Botrytis cinerea* (El Oirdi and Bouarab, 2007). A recent study involving *N. benthamiana* indicated silencing of *NbSGT1* compromised the protective effect of systemic acquired resistance-induced plants on neighbouring plants against bacteria wilt caused by *R. solanacearum* (Cheol Song et al., 2016). In the current study, the chlorosis and yellowing of the infection sites on leaves were greater in transgenic tobacco plants than in WT plants at 6 dpi. Additionally, there were substantially more concentration of bacterial wilt and scrab bacteria in the transgenic plants than in the WT plants, indicating that the overexpression of *CmSGT1* in tobacco adversely affects the resistance to bacterial wilt and scab, which differs from the effects of *CmSGT1* overexpression on the resistance to PM. Consequently, *SGT1* may function differently depending on the plant-pathogen combinations with diverse effectors and R proteins.

In conclusions, the results of this study indicate that the overexpression of *CmSGT1* may increase the PM resistance, but decrease bacterial wilt and scab resistance, in transgenic tobacco plants. Additionally, *CmSGT1* overexpression may improve PM resistance by enhancing HR-related cell death and H_2O_2 accumulation and upregulating the expression of SA-mediated defense-response genes. Further analyses of the *CmSGT1* gene may be useful for characterizing biotic stress signaling pathways and for the genetic engineering of novel pumpkin cultivars. The results described herein may be relevant for future biotechnology-based investigations and the molecular breeding of pumpkins and related plant species.

1.4 Materials and Methods

1.4.1 Plant materials and treatments

Pumpkin (*C. moschata*) inbred line 112-2 and cultivar 'Jiujiangjiaoding' (abbreviated 'JJJD'), which are resistant and susceptible to PM, respectively, were provided by the Henan Institute of Science and Technology, Xinxiang, Henan, China (Zhou et al., 2010). Pumpkin seeds were germinated and the resulting seedlings were grown as previously described (Guo et al., 2018). Seedlings at the three-leaf stage were treated as follows. The PM infection was initiated as described by Guo et al. (2018). Specifically, conidia were collected from the leaves of pumpkin plants naturally infected with PM in a local greenhouse. Seedlings were sprayed with a freshly prepared spore suspension (10^6 spores/mL) or an exogenous signaling molecule or hormone, including 1.5 mmol/L H_2O_2, 100 μmol/L SA, 100 μmol/L abscisic acid (ABA), 100 μmol/L MeJA, 0.5 g/L ETH. Moreover, water alone was used for the control treatment (CK). The treated seedlings were maintained in a growth chamber with a 15 h light (28℃) /9 h dark (18℃) cycle (5 500 lux light intensity) and harvested after 0, 3, 6, 12, 24, and 48 h to examine the *CmSGT1* expression pattern. At each time point, two young leaves were collected from the upper parts of four seedlings (i.e., one sample), wrapped in foil, frozen in liquid nitrogen, and stored at −80℃. The treatments were arranged in a randomized complete block design, with three biological replicates.

1.4.2 Isolation of *CmSGT1* cDNA clone and sequence analysis

The *SGT1* homolog expressed sequence tag (GenBank accession number SRR5369792) was identified in a PM-resistant pumpkin seedling transcriptome by Guo et al. (2018). The full-length open reading frame (ORF) of the *SGT1* homolog was obtained with the cDNA fragment of this homolog used as a probe, as previously described (Guo et al., 2014). The theoretical molecular weight (Mw) and isoelectric point (pI) were calculated with the ExPASy Compute pI/Mw tool (Bjellqvist et al., 1993). Sequence data were analyzed with the ClustalW program (Thompson et al., 1994). The NCBI databases were screened for homologous sequences with the default parameters of the BLAST program (http://www.ncbi.nlm.nih.gov/blast) (Altschul et al., 1997).

1.4.3 Subcellular localization analysis of CmSGT1

The *CmSGT1* ORF (without the termination codon) was ligated into the pBI221-GFP vector for the subsequent production of a green fluorescent protein (GFP) -tagged CmSGT1 fusion protein. Polyethylene glycol was used during the transformation of *Arabidopsis thaliana* protoplasts with the recombinant plasmid (Lee et al., 2013). The subcellular localization of CmSGT1 was determined based on the GFP signal, which was detected with the LSM 510 Meta confocal fluorescence microscope (Zeiss, Jena, Germany).

1.4.4 Generation of *CmSGT1*-overexpressing transgenic tobacco plants

Pumpkin are known to be one of the plants most refractory for transformation. To date, only two reports on transformation in *C. moschata* existed using a combined method of vacuum infiltration and *Agrobacterium* infection (Nanasato and Tabei, 2015; Nanasato, et al., 2011). So, we chose an heterologous overexpression assay in tobacco instead of generating *C. moschata* transgenic plants. Forward and reverse primers with an added *Bam*H I site and *Kpn* I site, respectively, were used to amplify *CmSGT1*. The amplified sequence was then inserted into the pVBG2307 vector for the subsequent expression of *CmSGT1* under the control of the 35S cauliflower mosaic virus (CaMV) promoter. The recombinant plasmid was introduced into *Agrobacterium tumefaciens* GV3101 cells as previously described (Guo et al., 2014). The resulting *A. tumefaciens* cells were used to transform tobacco (*Nicotiana benthamiana*) plants according to a previously described leaf disk method (Li et al., 2012). The transgenic tobacco plants were confirmed by examining the segregation ratio of the kanamycin selectable marker and by PCR analysis of *NPTII* and *CmSGT1*. T2 lines that produced 100% kanamycin-resistant plants in the T3 generation were considered as homozygous transformants. In each experiment, T2 generations of homozygous transgenic lines (L3, L5 and L11) were selected for further analysis. Similar phenotypes and results used for this study were observed in more than three independent lines of transgenic plants.

1.4.5 Primer design

Details regarding all of the primers designed and used in this study are provided in Table 2-2.

Table 2-2　Primers used in this investigation

Gene	Accession	Primer sequence (5′-3′)
CmSGT1	MH105820	F: CATCAGTTATCAAGACTTCCAAGTC R: GAGTTGAAGAAATGGGAGATCTGAT
RT-qPCR for CmSGT1		F: ATTACCCAGAGCATTAGTGTCCC R: TATCCTCGTCAGCGTCCTTGTAT
Over-expression vector		F: GGG<u>GGATCC</u>CATCAGTTATCAAGACTTCCAAGTC (BamH I) R: GGG<u>GGTACC</u>GAGTTGAAGAAATGGGAGATCTGAT (Kpn I)
NtNPR1	U76707	F: ACATCAGCGGAAGCAGTAG R: GTCGGCGAAGTAGTCAAAC
NtPR1a		F: CCTCGTACATTCTCATGGTCAAT R: CCATTGTTACACTGAACCCTAGC
NtPR5		F: CCGAGGTAATTGTGAGACTGGAG R: CCTGATTGGGTTGATTAAGTGCA
NtPDF1.2	T04323	F: GGAAATGGCAAACTCCATGCG R: ATCCTTCGGTCAGACAAACG
NtPAL	X95342	F: GTTATGCTCTTAGAACGTCGCCC R: CCGTGTAATGCCTTGTTTCTTGA
NtEF1-α	AF120093	F: TGTGATGTTTTTGTTCGGTCTTTAA R: TCAAAAGAAAATGCAGACAGACTCA
NbSGT1	AF516180	F: GGTTGTTTGGGAAGATAACACC R: ATTTCGAGGAAGGATAACTGGG
β-actin		F: TCTCTATGCCAGTGGTCGTA R: CCTCAGGACAACGGAATC
NPTII		F: AGACAATCGGCTGCTCTGAT R: TCATTTCGAACCCCAGAGTC

1.4.6 Analysis of transgenic tobacco plants infected with powdery mildew

Conidia were collected from tobacco leaves naturally infected with *Erysiphe cichoracearum* DC. , which causes PM. Transgenic tobacco seedlings at the five-leaf stage were used for phenotypic analyses. Specifically, the petiole of the second leaf from the top of the seedlings was wrapped with cotton moistened with water. The leaf was placed on a porcelain tray containing filter paper, after which it was sprayed with a spore suspension (10^6 spores/mL). The tray was then covered with film to maintain humidity. At 10 dpi, disease severity of leaves in vitro were calculated as $[(5A+4B+3C+2D+E)/5F] \times 100$ according to Ishii et al. (2001). Additionally, the first fully-expanded true leaf of the transgenic and wild-type plants were sprayed with the above-mentioned spore suspension and sampled after 0, 12, 24, 48, 72, and 120 hpi, frozen in liquid nitrogen, stored at $-80°C$ and used for extraction of total RNA. On the other hand, these leaves were harvested symmetrically along the sides of the main vein after 0, 1, 3, 4, 5, and 7 d to examine of cell death and H_2O_2 accumulation.

Leaves from the transgenic tobacco seedlings were stained with 3, 3′-diaminobenzidine (DAB) and trypan blue staining as previously described (Choi et al., 2012) to analyze H_2O_2 accumulation and cell death. For DAB staining, whole leaves were immersed in DAB solution (1 mg/mL, pH 5.7) and incubated overnight (almost 8 h) in darkness. Leaves were then de-stained three times with 95% ethanol. Regarding trypan blue staining, whole leaves were boiled in a lactophenol-ethanol trypan blue solution (10 mL lactic acid, 10 mL glycerol, 10 g phenol, 30 mL absolute ethanol, and 10 mg trypan blue dissolved in 10 mL distilled water) for 10 min and then maintained at room temperature overnight. Leaves were de-stained with 2.5 g/mL chloral hydrate in distilled water.

1.4.7 Analysis of transgenic tobacco plants infected with bacterial wilt and scab

The sixth and twelfth leaf from the top of transgenic tobacco seedlings were inoculated with bacterial solutions (10^8 cfu/mL). Specifically, the bacterial solutions were injected into the underside of leaves between the lateral veins with a syringe lacking a needle. Each leaf was inoculated in four places. The petiole of leaves placed on a porcelain plate was wrapped with water-saturated degreased cotton, after which the plate was covered with film to maintain humidity and then incubated at 28℃ with a 16 h light/8 h dark cycle. Water was periodically added to the cotton to maintain an appropriate moisture level. After a 6 day incubation, the concentration of bacteria at the injection sites was determined as follows. The injection sites were sampled with a circular perforator (1cm diameter) and ground in aseptic water (0.1 g added to 0.9 mL water). The ground material was serially diluted to produce the 10^2 fold, 10^3 fold, and 10^4 fold diluents. A 0.1-mL aliquot of the 10^3 fold, 10^4 fold, and 10^5 fold diluents were added to a petri dish containing NA medium. The medium was incubated upside down at 28℃. Plates with 30-300 colonies were considered ideal. The bacterial concentration was calculated with the following formula: colony average×10×dilution/g.

1.4.8 Quantitative real-time PCR analysis

The RNA extraction, first-strand cDNA synthesis and quantitative real-time (qRT-PCR) were completed as described by Guo et al. (2015). Relative gene expression levels were determined with the $2^{-\Delta\Delta CT}$ method. Total RNA was extracted from the leaves of pumpkin seedlings treated with various stresses or distilled water for 0, 3, 6, 12, 24, or 48 h as described above. The β-actin gene was used as an internal control, because it was confirmed as a suitable reference gene for normalizing of gene expression levels in pumpkin (Wu and Cao, 2010).

For signal-related genes' expression treatment, the first fully-expanded true leaf were collected from both PM-treated and control plants after 0, 12, 24, 48, 72, and 120 hpi. At each time point, sample was frozen in liquid nitrogen, stored at $-80°C$ and used for extraction of total RNA.

Total RNA was extracted from *CmSGT1*-overexpressing transgenic and WT tobacco seedlings to examine the expression of five hormone-related genes (*NtNPR1*, *NtPR1a*, *NtPR5*, *NtPDF1.2* and *NtPAL*) and tobacco isoform (*NbSGT1*). The tobacco *NtEF1-α* gene (AF120093) was included in the assays as an internal control.

1.4.9 Statistical analyses

Values were expressed as the mean ± standard error of three independent determinations. Data were compared by analysis of variance (ANOVA) using an one-way ANOVA, and differences between WT and transgenic plants were tested by a post hoc comparison test (Student-Newman-Keuls) at $P < 0.05$ with SPSS 19.0 for Windows (SPSS Inc, Chicago, IL).

2 Expression of pumpkin *CmbHLH87* gene improves powdery mildew resistance in tobacco

Powdery mildew (PM), caused by *Podosphaera xanthii*, is a major threat to the global cucurbit yield. The molecular mechanisms underlying the PM resistance of pumpkin are largely unknown. A homolog of the basic helix-loop-helix (bHLH) transcription factor was previously identified through a transcriptomic analysis of a PM-resistant pumpkin. In this study, this bHLH homolog in pumpkin has been functionally characterized. *CmbHLH87* is present in the nucleus. CmbHLH87 expression in the PM-resistant material was considerably downregulated by PM; and abscisic acid, methyl jasmonate, ethephon, and NaCl treatments induced CmbHLH87 expression. Ectopic expression of *CmbHLH87* in tobacco plants alleviated the PM symptoms on the leaves, accelerated cell necrosis, and enhanced H_2O_2 accumulation. The expression levels of *PR1a*, *PR5*, and *NPR1* were higher in the PM-infected transgenic plants than in PM-infected wild-type plants. Additionally, the chlorosis and yellowing of plant materials were less extensive and the concentration of bacteria at infection sites was lower in the transgenic tobacco plants than in the wild-type plants in response to bacterial wilt and scab pathogens. *CmbHLH87* may be useful for genetic engineering of novel pumpkin cultivars in the future.

2.1 Introduction

Pumpkin is an important vegetable crop and is widely cultivated in China, with total harvested area of 438 466 hectares in 2017 (i.e., 17.42% of the global area) (Food and Agriculture Organization, 2017). Powdery mildew (PM) is a fungal disease seriously affecting Cucurbitaceae crops yield including cucumber, melon, watermelon, pumpkin and squash in the world. The disease is mainly caused by *Podosphaera xanthii* (formerly known as *Sphaerotheca fuliginea*), which is a biotrophic plant pathogen (Fukino et al., 2013; Perez-Garcia et al., 2009). Excessive fungicide application poorly control PM, because it not only increases selection pressure on PM pathogens to adapt increasing levels of fungicide resistance, but it also may be harmful for human health and the environment. Therefore, studying the moleculer mechanism of PM by exploiting the resistant genes to breed resistant varieties represents a favored strategy to control PM.

The basic helix-loop-helix (bHLH) transcription factors constitute one of the largest transcription factor families in plants, wherein they help regulate developmental processes and responses to environmental stresses. These proteins have a 60-amino-acid conserved domain, which contains the following two functionally distinct regions: an N-terminal basic region (13-17 amino acids) that functions as a DNA-binding domain and a C-terminal HLH region that contributes to the formation of homodimers or heterodimers (Heim et al., 2003; Toledo-Ortiz et al., 2003). Recent studies have indicated that a number of bHLH transcription factor genes are involved in responses to abiotic stresses including drought, salt, and cold. The overexpression of *AtbHLH068* and *OsbHLH148* in transgenic *Arabidopsis thaliana* and rice, respectively, reportedly induces drought stress tolerance via abscisic acid (ABA) - and jasmonic acid (JA) -mediated signaling pathways (Le Hir et al., 2017; Seo et al., 2011). In rice, *OsbHLH062*, *OsJAZ9*, and *OsNINJA* form a transcriptional regulatory complex that fine-tunes the expression of JA-responsive genes involved in salt stress tolerance (e.g., *OsHAK21*) (Wu et al., 2015). Additionally, *VaICE1*/*VaICE2*, *ZmmICE1*, and *NtbHLH123* are key regulators in the C-repeat binding factor regulatory pathway controlling cold tolerance (Xu et al., 2014; Lu et al., 2017; Zhao et al., 2018). Moreover, bHLHs influence the adaptation and resistance of plants to pathogen stress. An earlier investigation revealed that *OsDPF* expression is induced in the leaves of blast-infected rice plants, and the overexpression and knockdown of *DPF* considerably increases and decreases the accumulation of momilactones and phytocassanes, respectively (Yamamura et al., 2015). The overexpression of wheat *bHLH060* in transgenic *A. thaliana* negatively regulates the resistance to *Pseudomonas syringae* through JA and ethylene (ETH) signaling pathways (Wang et al., 2015). A recent study on wheat indicated that the expression levels of 28 and 6 *TabHLH* genes are obviously downregulated and upregulated, respectively, in response to a powdery mildew infection (Wei and Chen, 2018). However, bHLH functions related to biotic stress resistance remain poorly characterized in plants.

In our previous study, the PM-resitant pumpkin resources were obtained in an 8 year outdoor field observation study (Zhou et al., 2010). However, the resistant mechanism underlying pumpkin responseto biotic stress is not yet elucidated. Thus, a RNA sequencing analysis of PM-infected pumpkin identified 4 716 differentially expressed genes, including gene encoding bHLH transcription factor (*bHLH87*) (Guo et al., 2018). The expression of this *bHLH87* homolog in response to PM, hydrogen peroxide (H_2O_2), salicylic acid (SA), abscisic acid (ABA), methyl jasmonate (MeJA), ethephon (ETH), and NaCl treatments were analyzed through realtime quantitative PCR (RT-qPCR). To evaluate its function in disease resistance, we expressed *CmbHLH87* in tobacco ectopically. Transgenic tobacco plants constitutively overexpressing *CmbHLH87* were more resistant to PM, bacterial wilt, and scab than the control plants.

2.2 Results

2.2.1 Isolation of *CmbHLH87* and subcellular localization

Pumpkin PM-related genes were identified in a transcriptome

(Guo et al., 2018). One of these clones was 89% identical at the nucleotide level to *C. melo bHLH87*. Full-length of this homolog was obtained by a homology-based candidate gene method (named *CmbHLH87*) and submitted to the GenBank database (accession number MH105822). The size of this gene was 1 222 bp, including a 1 068 bp open reading frame (355 amino acids). The predicted polypeptide was relatively acidic, with a pI of 5.96, and a Mw of 39.1 ku. A phylogenetic tree was conducted between the overall amino acid sequences of CmbHLH87 and other known bHLH proteins. CmbHLH87 was clustered into the bHLH Ⅷ subfamily (Heim et al., 2003) (Fig. 2-8). An alignment of the deduced CmbHLH87 amino acid sequence with homologous sequences is presented in Fig. 2-9. At the amino acid level, CmbHLH87 was highly similar to the bHLH87 transcription factors from *C. moschata* (99.1% identical), *Cucurbita pepo* (98.8% identical), and *Cucurbita maxima* (71.0% identical), but was relatively dissimilar to *Nicotiana tabacum* bHLH87 (38.3% identical) and *Arabidopsis* AtbHLH087 (41.7% identical). The predicted amino acid sequence contained a conserved bHLH domain (amino acids 256-305) and a helix-loop-helix structure at the C terminal.

The subcellular localization of CmbHLH87 was assessed with a CmbHLH87-GFP fusion protein into *Arabidopsis* protoplasts under the control of the 35S CaMV promoter. The GFP signal in protoplasts producing GFP alone was detected in the cytoplasm and nucleus (Fig. 2-10) supported by Guo et al. (2019), whereas the signal from the CmbHLH87-GFP fusion protein was detected exclusively in the nucleus.

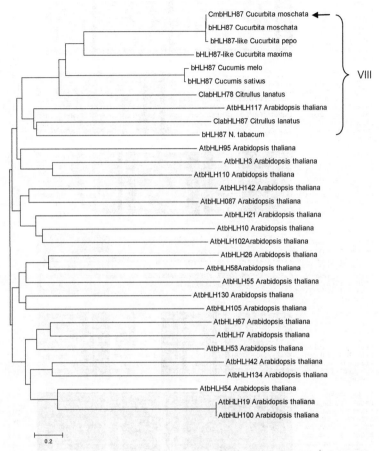

Fig. 2-8　A Neighbor-joining phylogenetic tree of CmbHLH87 and bHLH proteins from different plant species. The CmbHLH87 protein is marked with arrow. The bHLH subgroup names are shown to the right of square.

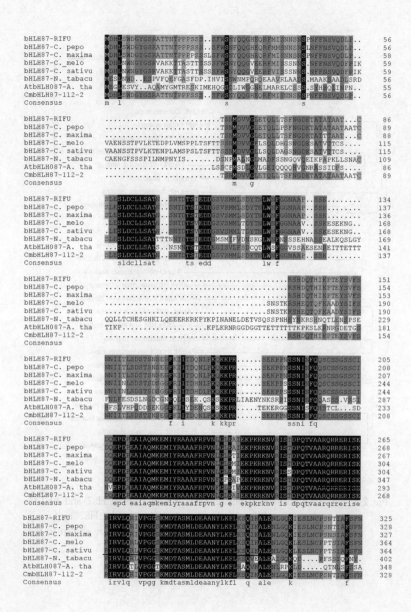

第 2 章 南瓜抗白粉病相关基因的分离及功能研究

```
bHLH87-RIFU       ..FFPIQTSSSSSSHNNFSLQNEYDQLM.....     352
bHLH87-C.pepo     ..FFPIQTSSSSSHNNFSLQNEYDQLM.....      355
bHLH87-C.maxima   P.FFPIQTSSSSSHNNFSLQNEYDQLM.....      354
bHLH87-C._melo    P.FFPIQTSS...SHNNETLLNPNHLINQYPQN     393
bHLH87-C._sativu  P.FFPIQTSS...SHNNETLLNPNHLINQYPQN     393
bHLH87-N._tabacu  ..PFPMQLPH......EPVCNPNPIHGPKS..      424
AtbHLH087-A.tha   PTSFPLFHPE.....FLPLQNPNQIHHPEC..      373
CmbHLH87-112-2    ..FFPIQTSSSSSHNNFSLQNEYDQLM.....      355
Consensus            fp             np
```

Fig. 2-9 Amino acid sequences alignment of pumpkin CmbHLH87 with others. The conserved domains (bHLH) are shown by thin underlines. The genes included are bHLH87-like (*C. moschata* cv. 'Rifu', XP _ 022953326.1), bHLH87-like (*C. pepo*, XP _ 023547764.1), bHLH87-like (*C. maxima*, XP _ 022992322.1), bHLH87-like (*C. melo*, XP _ 008462383.1), bHLH87-like (*C. sativus*, XP _ 004141799.1), bHLH87-like (*Nicotiana tabacum*, XP _ 016504990.1), AtbHLH087 (*Arabidopsis*, AAM10960.1) and CmbHLH87 (*C. moschata* cv. '112-2', MH105822).

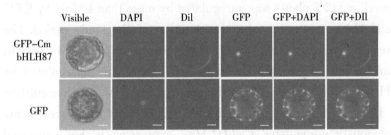

Fig. 2-10 The subcellular localization of pumpkin CmbHLH87. The fused green fluorescent protein-CmbHLH87 GFP and alone GFP constructs were introduced into Arabidopsis protoplast by polyethylene glycol-mediated protoplast transformation. The fluorescent signals were detected using a confocal fluorescence microscope. Scale bars=5 μm.

2.2.2 Expression patterns of *CmbHLH87* in response to PM and exogenous treatments

The *CmbHLH87* expression patterns in both PM-resistant inbred line 112-2 and PM-susceptible cultivar 'JJJD' were analyzed in response to PM and exogenous treatments (H_2O_2, SA, ABA, MeJA, ETH, or NaCl) (Fig. 2-11). The *CmbHLH87* expression level in the 112-2 plants was significantly downregulated by PM (except at 0 hpi), but was significantly upregulated by H_2O_2 treatment (except at 6 hpi). In contrast, the *CmbHLH87* expression level in 'JJJD' plants was not significantly altered by PM, but was downregulated by H_2O_2 treatments. Regarding the effects of SA, the *CmbHLH87* expression level in 112-2 plants decreased at 3 and 6 hpi, and was essentially unchanged thereafter. However, the *CmbHLH87* expression level in 'JJJD' plants was significantly upregulated by the SA treatment. In response to ABA, MeJA, ETH, and NaCl treatments, *CmbHLH87* expression was significantly higher in the 112-2 and 'JJJD' seedlings than in the control (CK) seedlings. Specifically, the *CmbHLH87* expression level in 112-2 plants was upregulated by more than 40 fold by ETH and NaCl treatments over the duration of the study period. The results indicated that *CmbHLH87* expression in PM-resistant inbred line 112-2 was downregulated by PM and upregulated by H_2O_2, which differed from the results of the PM-sensitive cultivar 'JJJD'. Moreover, ABA, MeJA, ETH, and NaCl treatments obviously upregulated *CmbHLH87* expression in both analyzed plant materials and these were not influenced by susceptibility or resistance status.

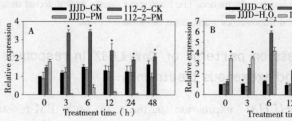

第2章 南瓜抗白粉病相关基因的分离及功能研究

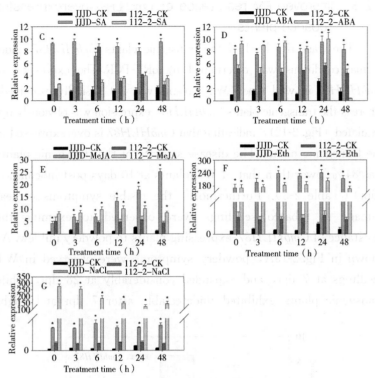

Fig. 2-11　*CmbHLH87* expression in response to powdery mildew and exogenous treatments. The pumpkin seedlings were sprayed with a spore suspension (A), exogenous hydrogen peroxide (H_2O_2) (B), salicylic acid (SA)(C), abscisic acid (ABA)(D), methyl jasmonate (MeJA)(E), ethephon (ETH)(F), and NaCl (G). The β-actin gene was used as an internal reference for qRT-PCR. The transcript level of CmbHLH87 in the cultivar 'JJJD' at 0 h was used as control (quantities of calibrator) and assumed as 1. The values are the means±SEs of three biological replicates. Data between treatments (112-2-treatment vs. 112-2-CK and JJJD-treatment vs. JJJD-CK) were analyzed by one-way ANOVA, and * denotes statistical significance at $P<0.05$.

2.2.3 Improved PM resistance of *CmbHLH87*-overexpressing tobacco plants

The transcript level of the transgenic plants *CmbHLH87* under normal conditions was determined by qRT-PCR. The expression of *CmbHLH87* in wild type (WT) plants was not basically examined, whereas the transgenic plants *CmbHLH87* expression was obviously up-regulated (Fig. 2-12), indicating that *CmbHLH87* is overexpressed in the transgenic plants. The disease severity of the transgenic plants was 84% lower than that of WT plants at 10 days post-inoculation (dpi) (Table 2-3). Furthermore, the visible symptoms of leaf damage in tobacco seedlings were observed to examine the resistance of *CmbHLH87*-expressing plants to powdery mildew. As shown in Fig. 2-13A, powdery symptoms were observed in WT seedlings at 7 dpi, and expanded considerably at 28 dpi, while transgenic plants exhibited undetectable after 7 dpi and slight

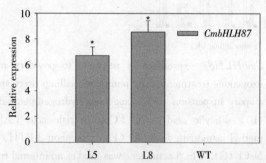

Fig. 2-12 Relative expression of *CmbHLH87* in transgenic or wild-type plants under normal growth conditions. L5 and L8 are two independent transgenic lines that overexpress *CmbHLH87*. Three biological triplicates were averaged and bars indicate standard error of the mean. * denotes statistical significance at $P < 0.05$ between both materials.

powdery pots at 28 dpi. Blue spots are actually trypan blue stained cell death. The blue spots on the transgenic leaves ocurred at 4 dpi, continued to expand at 5 and 7 dpi, and those blue spots were more and bigger than those on the WT leaves, implying that overexpression of *CmbHLH87* in transgenic plants accelerated cell death following a PM infection (Fig. 2-13). Moreover, brown spots are H_2O_2 for DAB staining. Compared with WT leaves, there were more brown spots on the transgenic leaves at 1 dpi, larger and stained more intensely at 3 and 5 dpi. These results indicated that the overexpression of *CmbHLH87* promoted the accumulation of H_2O_2 in transgenic plants infected with PM (Fig. 2-13C).

Table 2-3 Disease severity of leaves of tobacco seedlings infected with powdery mildew

Materials	Disease severity (in vitro leaf)
L5	6.50±1.03
L8	7.60±1.01
L12	8.00±1.21
WT	45.80±1.41*

Note: Data are mean values (±SD) of at least three independent experiments. * indicates significantly different values between treatments ($P<0.05$). L5, L8, and L12 are three independent transgenic lines that overexpress *CmbHLH87*.

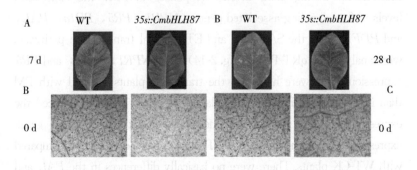

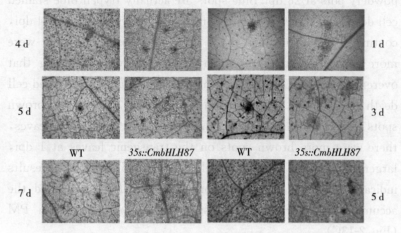

Fig. 2-13 The pathogenic symptoms, trypan blue, and diaminobenzidine (DAB) staining of tobacco leaves infected with powdery mildew. The pathogenic symptoms of transgenic tobacco (L5) and wild type (WT) at 7 and 28 day post inoculation (A); trypan blue staining was performed to visualize cell death (B); DAB staining was performed to visualize hydrogen peroxide (H_2O_2) accumulation (C). Scale bars=200 mm.

2.2.4 Expression of signal-related genes in transgenic tobacco plants

To investigate the signal transduction pathways affected by CmbHLH87 during plant defense responses to PM, the expression levels of five signaling-associated genes (*NPR1*, *PR5*, *PR1a*, *PAL*, and *PDF1.2*) in the SA, JA, and ETH signal transduction pathways were analyzed by qRT-PCR (Fig. 2-14). The *NPR1*, *PR1a*, and *PR5* expression levels were higher in the transgenic plants infected with PM than in the transgenic plants-CK plants, implying that PM induced the expression of these genes. And the opposite pattern of these genes expression was observed in the WT plants infected with PM compared with WT-CK plants. There were no basically differences in the *PAL* and

第2章 南瓜抗白粉病相关基因的分离及功能研究

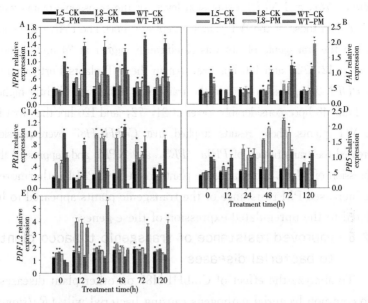

Fig. 2-14 Expression of signal-related genes in transgenic and wild-type (WT) plants treated with powdery mildew. The samples of two genetically-modified tobacco lines (L5, L8) were used to analyze genes expression by qRT-PCR. A, *NtNPR1*; B, *NtPAL*; C, *NtPR1a*; D, *NtPR5*; E, *NtPDF1.2*. There were four treatments: L5/L8-CK represents transgenic lines under normal conditions; L5/L8-PM represents transgenic lines infected with powdery mildew; WT-CK represents WT plants under normal conditions; WT-PM represents WT plants infected with powdery mildew. Tobacco *NtEF1-α* gene (AF120093) was used as an internal reference. The expression levels of signal-related genes in WT plants at 0 h were used as control (quantities of calibrator) and assumed as 1. Three biological triplicates per line were averaged and Bars indicate standard error of the mean. Data between transgenic lines and WT plants (L5/L8-PM vs. WT-PM and L5/L8-CK vs. WT-CK) were analyzed by one-way ANOVA, and * denotes statistical significance at $P<0.05$.

PDF1.2 (except at 12 hpi) expression levels of the transgenic plants between PM and CK treatments. In response to the PM infection, the

PR1a, *PR5*, and *NPR1* expression levels of the transgenic plants were higher than those of the WT plants(except at 0 hpi). The *PAL* expression level in the transgenic plants was significantly higher at 24 and 48 hpi, lower at 12, 72, and 120 hpi than that in the WT plants. Furthermore, the *PDF1.2* expression level in the transgenic plants was slightly higher at 12 and 48 hpi, considerably lower at 24, 72, and 120 hpi than that in the WT plants. These results implied that *CmbHLH87* overexpression increased the transcription of *PR1a*, *PR5*, and *NPR1* and suppressed the expression of *PDF1.2* in transgenic plants infect with PM. Furthermore, the increased PM resistance of the transgenic plants appeared to be related to the upregulated expression of these genes.

2.2.5 Improved resistance of transgenic tobacco plants to bacterial diseases

To analyze the effect of CmbHLH87 on other plant diseases, two common bacterial pathogens causing bacterial wilt (*Ralstonia solanacearum*) and scab (*Xanthomonas euvesicatoria*) were injected into tobacco plants (Fig. 2-15). At 6 dpi, the chlorosis and yellowing of the 6th leaf at injection sites were less severe for the transgenic plants than for the WT plants. Additionally, the concentrations of bacterial wilt and scab bacteria in the transgenic plants were 0.14 and 0.10 times those of the WT plants, respectively. These observations suggested that overexpression of *CmbHLH87* in tobacco increased the resistance to bacterial wilt and scab.

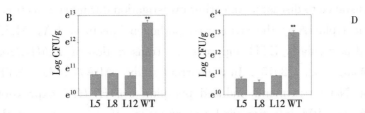

Fig. 2-15　The resistance of CmbHLH87 in tobacco plants to bacterial wilt and scab. (A) Pathogens symptoms of the 6th-upper leaf injection sites was injected with bacterial wilt bacteria with a needle-removed syringe; (B) concentration bacteria of the 6th-upper leaf injection sites was injected with bacterial wilt bacteria with a needle-removed syringe; (C) pathogens symptoms of the 6th-upper leaf injection sites was injected with scab bacteria with a needle-removed syringe; (D) concentration bacteria of the 6th-upper leaf injection sites was injected with scab bacteria. Three biological triplicates were averaged and bars indicate standard error of the mean. ✳✳denotes significant differences between wild-type (WT) and transgenic plants at $P<0.01$.

2.3　Discussion

　　In this study, we isolated a novel pumpkin *bHLH87* gene and the predicted amino acid sequence included a conserved bHLH domain, suggesting that CmbHLH87 is a putative pumpkin bHLH transcription factor. Its predicted pI was 5.96, which is similar to a previous report that the isoelectric points of watermelon bHLH family proteins are mostly in the acidic range (He et al., 2016). The CmbHLH87 protein was localized to the nucleus in *Arabidopsis* protoplasts, which is consistent with the observations of earlier studies involving the bHLH transcription factors from other plant species (Wang et al., 2015).

　　The interplay among complex signaling networks, including various pathways regulated by phytohormones, such as SA, JA, ethylene (ETH), and ABA, considerably influences plant resistance to diseases. An earlier investigation during an infection of wheat plants by the stripe rust pathogen infection, SA, MeJA, and particularly ETH suppress the transcription of *TabHLH060* (Wang et al., 2015). In the current study, ABA, MeJA, ETH, and NaCl treatments induced pumpkin *CmbHLH87* expression, whereas PM infection had the opposite effect, suggesting that *CmbHLH87* may have a regulatory role during responses to hormones, salt stress, and PM. Additionally, overexpression of the pumpkin *CmbHLH87* gene in tobacco plants decreased the disease severity by 84%, accelerated necrosis, and increased the accumulation of H_2O_2. These results indicated that the PM resistance of the transgenic tobacco plants was enhanced, likely because of the changes to the HR-related necrosis and H_2O_2 accumulation. Our findings are consistent with the results of an earlier investigation that H_2O_2 accumulation and subsequent cell death usually lead to the resistance to diseases caused by biotrophic pathogens (Li et al., 2011). The expression pattern of *CmbHLH87* in response to PM is controversial to its transgenic phenotype in disease resistance. There were similar reports that pathogen-regulated ectopic expression of bHLH transcription factors were inconsistent with activation of pathogen resistance. Two *Arabidopsis* bHLH25 and bHLH27 positively influence cyst nematode parasitism (*Heterodera schachtii*). Transgenic *Arabidopsis* plants overexpressing either one or both of the bHLH genes exhibited an increased susceptibility to *H. schachtii* (Jin et al., 2011). Jin et al. document an example of pathogen-induced ectopic co-expression of two regulatory genes to enhance pathogen success. In our experiment, This is also an intriguing biological phenomenon that highlights the complexity of obligate biotrophic plant-pathogen interactions. who are direct target of the CmbHLH87 protein and how to interact in response to pathogen resistance need to be further research. Powdery mildew symptoms herein were observed in WT tobacco seedlings at 7 dpi of *G. cichoracearum*, and the phenotypes were not more robust than those of previous reports: powdery mildew disease occurs abundantly in tobacco plants at 12 dpi of the powdery mildew strain *G. cichoracearum* SICAU1 (Li et al., 2019). Different tobacco accessions may show various susceptible powdery mildew disease phenotypes to different powdery mildew isolates.

SA plays a central role in plant disease resistance against biotrophic pathogens, while JA is critical for activating plant defenses against necrotrophic pathogens (Wasternack et al., 2013). *PDF1.2* is important for the JA/ETH-dependent signaling pathway. At intermediate SA levels, NPR1 (non-expresser of PR1) accumulates and interacts with the TGA transcription factor, functioning as a co-activator of SA-responsive genes, including *PR* genes (Caarls et al., 2015). The overexpression of wheat *TabHLH060* gene in *Arabidopsis* increases the susceptibility to *P. syringae* by suppressing the transcription levels of *PR1*, *PR2* and *PR5*, which are involved in the SA signaling pathway, and by upregulating the expression of *PDF1.2* and *ORA59*, which are associated with the JA and ETH signaling pathways (Wang et al., 2015). In bread wheat, *TaJAZ1* overexpression increases PM resistance by promoting the accumulation of reactive oxygen species. The encoded TaJAZ1 directly interacts with TaMYC4 (JA-induced bHLH transcription factor) to repress its transcriptional activity (Jing et al., 2019). In the current study, following PM infection, the *PR1a*, *PR5*, and *NPR1* expression levels in the transgenic plants were higher than those in the WT plants, whereas a different expression pattern was observed for the *PDF1.2* expression level. These results suggest that in the SA pathway, the transactivation of *PR1a*, *PR5*, and *NPR1* is dependent on CmbHLH87. Additionally, CmbHLH87 does not directly affect the JA/ETH-dependent defense pathway to regulate *PDF1.2* expression. We propose that CmbHLH87 activates stress-resistance mechanisms via SA-dependent defense pathways without upregulating *PAL* expression, but suppresses the activities of JA/ETH-dependent defense pathways. Therefore, we speculate that CmbHLH87 positively regulates the H_2O_2 and SA pathways. Moreover, H_2O_2 might directly transfer the SA signal to regulate the expression of downstream response genes in the *CmbHLH87*-overexpressing transgenic plants infected with PM. In NPR1-dependent SA signal transduction pathways, the activation of *PR* gene expression requires an interaction between NPR1 and the TGA transcription factor that binds to the target promoters (Spoel and Dong, 2012; Fan and Dong, 2002). We speculated whether there is a relationship between the upregulation of *NPR1* expression and SA-mediated transcriptional activation of *PR* genes. Notably, the phenotypes and genes influenced by *CmbHLH87* overexpression in tobacco plants might not be the same as those regulated by *CmbHLH87* expression in pumpkin in response to PM. Additional studies are necessary to reveal the biological functions of CmbHLH87 in pumpkin infected with PM.

Two globally important diseases that affect tobacco, bacterial wilt and scab, are caused by the necrotrophic *R. solanacearum* and the hemi-biotrophic *X. euvesicatoria* respectively. The overexpression of the bHLH transcription factor gene (*HBI1*) decreases the pathogen-associated defense responses and increases the susceptibility to bacteria *P. syringae* (Malinovsky et al., 2014).

In the current study, the chlorosis and yellowing of the leaves near infection sites were less extensive in the transgenic plants than in the WT plants at 6 dpi. Moreover, the concentrations of bacterial wilt and scab bacteria were substantially lower in the transgenic plants than in the WT plants. These results imply overexpression of the pumpkin *CmbHLH87* gene in tobacco enhances the resistance to bacterial wilt and scab, which is consistent with the effects of *CmbHLH87* overexpression on the resistance to PM.

In conclusions, the results of this study indicate that overexpression of the pumpkin *CmbHLH87* gene in tobacco may increase the resistance to PM, bacterial wilt, and scab. Additionally, *CmbHLH87* overexpression may improve PM resistance by enhancing HR-related cell death and H_2O_2 accumulation and by upregulating the expression of SA-mediated defense genes. The data generated in this study may provide valuable genetic information for the breeding of new disease-resistant pumpkin varieties.

2.4 Materials and Methods

2.4.1 Plant materials and treatments

Pumpkin inbred line 112-2 and cultivar 'Jiujiangjiaoding' (abbreviated 'JJJD'), which are resistant and susceptible to PM, respectively, were provided by the Henan Institute of Science and Technology, Xinxiang, Henan, China (Zhou et al., 2010). Pumpkin seeds were germinated and the resulting seedlings were grown as previously described (Guo et al., 2018). Seedlings at the third-leaf stage were treated as previously described by Guo et al. (2019). Seedlings were sprayed with a freshly prepared spore suspension (10^6 spores/mL), exogenous 1.5 mmol/L H_2O_2, 100 μmol/L SA, 100 μmol/L ABA, 100 μmol/L MeJA, 0.5 g/

L ETH and 0.4mmol/L NaCl. Moreover, water alone was used for the control treatment (CK). The treated seedlings were maintained in a growth chamber with a 15 h light (28℃) /9 h dark (18℃) cycle (5 500 lux light intensity) and harvested after 0, 3, 6, 12, 24, and 48 h to examine the *CmbHLH87* expression pattern. At each time point, two young leaves were collected from the upper parts of four seedlings (i.e., one sample), wrapped in foil, frozen in liquid nitrogen, and stored at −80℃. The treatments were arranged in a randomized complete block design, with three biological replicates.

2.4.2 Isolation of *CmbHLH87* and sequence analysis

The *bHLH* homolog expressed sequence tag (GenBank accession number SRR5369792) was identified from a transcriptome of PM-resistant pumpkin seedling (Guo et al., 2018). Full-length of this homolog was obtained using a homology-based candidate gene method (Guo et al., 2014). The theoretical molecular weight (Mw) and isoelectric point (pI) were calculated with the ExPASy Compute pI/Mw tool (Bjellqvist et al., 1993). Sequence data were analyzed with the ClustalW program (Thompson et al., 1994). The phylogenetic tree was constructed using Mega5.0 by the neighbor-joining method. The NCBI databases were screened for homologous sequences with the default parameters of the BLAST program (http://www.ncbi.nlm.nih.gov/blast) (Altschul et al., 1997).

2.4.3 Subcellular localization analysis of CmbHLH87

The *CmbHLH87* ORF (without the termination codon) was ligated into the pBI221-GFP vector for the subsequent production of a green fluorescent protein (GFP) -tagged CmbHLH87 fusion protein. Polyethylene glycol was used during the transformation of *Arabidopsis thaliana* protoplasts with the recombinant plasmid

(Lee et al., 2013). The subcellular localization of CmbHLH87 was determined based on the GFP signal, which was detected with the confocal fluorescence microscope (UltraVIEW VoX, Olympus, Japan) (Guo et al., 2019).

2.4.4 Generation of CmbHLH87 transgenic tobacco plants

Pumpkin are known to be one of the plants most refractory for transformation. To date, only two reports on transformation in pumpkin existed using a combined method of vacuum infiltration and *Agrobacterium* infection (Nanasato and Tabei, 2015; Nanasato et al., 2011). So, we choosed an ectopic overexpression assay in tobacco instead of generating pumpkin transgenic plants. Forward and reverse primers with an added *Bam*H I site and *Kpn* I site, respectively, were used to amplify *CmbHLH87*. The amplified sequence was then inserted into the pVBG2307 vector for the subsequent expression of *CmbHLH87* under the control of the 35S cauliflower mosaic virus (CaMV) promoter. The recombinant plasmid was introduced into *Agrobacterium tumefaciens* GV3101 cells as previously described (Guo et al., 2014). The resulting *A. tumefaciens* cells were used to transform tobacco (*Nicotiana tabacum* L. cv. NC89) plants according to a previously described leaf disk method (Li et al., 2012). The transgenic plants were confirmed by examining the segregation ratio of the kanamycin selectable marker and by PCR analysis of *NPTII* and *CmbHLH87*, self-pollinated to obtain homozygous T2 offspring. T2 lines that produced 100% kanamycin-resistant plants in the T3 generation were considered as homozygous transformants. In each experiment, T2 generations of homozygous transgenic lines (L5, L8 and L12) were selected for further analysis. Similar phenotypes and results used for this study were observed in more than three independent lines of transgenic plants.

2.4.5 Primer design

All primers designed and used in this study are provided in Table 2-4.

Table 2-4 Primers used in this investigation

Gene	Accession	Primer sequence (5'-3')
CmbHLH87	MH105822	F: TTTGTCATCGGTTCGGTAAG
		R: GACTCCCGTTCGTTCTCAAG
RT-qPCR for		F: CAAACTCATATCAAACCAACAGA
Cm bHLH87		R: TATCATCTCCTTCATTTGTGCTA
Over-expression		F: GGGGATCCTTTGTCATCGGTTCGGTAAG (BamH I)
vector		R: GGGGTACCGACTCCCGTTCGTTCTCAAG (Kpn I)
NtNPR1	U76707	F: ACATCAGCGGAAGCAGTAG
		R: GTCGGCGAAGTAGTCAAAC
NtPR1a		F: CCTCGTACATTCTCATGGTCAAT
		R: CCATTGTTACACTGAACCCTAGC
NtPR5		F: CCGAGGTAATTGTGAGACTGGAG
		R: CCTGATTGGGTTGATTAAGTGCA
NtPDF1.2	T04323	F: GGAAATGGCAAACTCCATGCG
		R: ATCCTTCGGTCAGACAAACG
NtPAL	X95342	F: GTTATGCTCTTAGAACGTCGCCC
		R: CCGTGTAATGCCTTGTTTCTTGA
NtEF1-α	AF120093	F: TGTGATGTTTTTGTTCGGTCTTTAA
		R: TCAAAAGAAAATGCAGACAGACTCA
β-actin		F: TCTCTATGCCAGTGGTCGTA
		R: CCTCAGGACAACGGAATC
NPTII		F: AGACAATCGGCTGCTCTGAT
		R: TCATTTCGAACCCCAGAGTC

2.4.6 Performance of transgenic lines infected with PM, bacterial wilt and scab

Conidia were collected from tobacco leaves naturally infected with *Golovinomyces cichoracearum*, which are main pathogen isolates of PM. The upper second leaf from the transgenic and WT seedlings at the fifth-leaf stage was sprayed with a spore suspension

(10^6 spores/mL) and contine to in vitro grow for 10 d (Guo et al., 2019). At 10 dpi, mildew development on each leaf disk was recorded, using the following scale: 0 = no visible mildew development, 1=0 to 5%, 2=6% to 25%, 3=26% to 50%, 4 =51% to 75%, and 5 = >76% of disk surface covered with mildew, as described by Ishii et al. (2001). Disease severity (DS) was calculated as [(5A+4B+3C+2D+E)/5F] ×100, where A, B, C, D, and E were the number of leaf disks corresponding to the scales, 5, 4, 3, 2, and 1, respectively, and F was the total number of leaf disks assessed. Additionally, the second leaf of the transgenic and WT plants was sprayed with the above-mentioned spore suspension and sampled at 0, 12, 24, 48, 72, and 120 hpi for a subsequent extraction of total RNA. Furthermore, these leaves were harvested symmetrically along the sides of the main vein after 0, 1, 3, 4, 5, and 7 d to examine cell death and H_2O_2 accumulation. 3, 3′-diaminobenzidine (DAB) and trypan blue staining were used to analyze H_2O_2 accumulation and cell death, respectively, as previously described (Choi et al., 2012). The treatments were arranged in a randomized complete block design with three replicates.

The upper sixth leaf from the transgenic and WT seedlings at the twelfth-leaf stage were inoculated with bacterial solutions (10^8 CFU/mL). The bacterial solutions were injected into the underside of leaves between the lateral veins with a syringe lacking a needle. After 6 d, the concentration of the bacteria at the injection sites was determined as previously described (Guo et al., 2019). The injection sites were sampled with a circular perforator (1cm diameter) and ground in aseptic water, then serially diluted to produce the 10^2 fold, 10^3 fold, and 10^4 fold diluents. Experiments were done in triplicate for each line.

2.4.7 qRT-PCR analysis

The RNA extraction, first-strand cDNA synthesis and qRT-PCR were completed as described by Guo et al. (2015). Relative gene expression levels were determined with the $2^{-\Delta\Delta CT}$ method. Total RNA was extracted from the leaves of pumpkin seedlings treated with various stresses or distilled water for 0, 3, 6, 12, 24, or 48 h as described above. The *β-actin* gene was used as an internal reference for normalizing of gene expression levels in pumpkin (Wu and Cao, 2010).

Total RNA was extracted from *CmbHLH87*-overexpressing and WT tobacco seedlings to examine the expression of five hormone-related genes (*NtNPR1*, *NtPR1a*, *NtPR5*, *NtPDF1.2*, and *NtPAL*) at 0, 12, 24, 48, 72, and 120 hpi as described above. The tobacco *NtEF1-α* gene (AF120093) was used as an internal control in the assays.

2.4.8 Statistical analyses

Values are herein provided as the mean ± standard error of three independent analyses. Data underwent a one-way analysis of variance, and differences between WT and transgenic plants were evaluated with a *post hoc* comparison test (Student-Newman-Keuls method) at $P < 0.05$ with SPSS 19.0 for Windows (SPSS Inc, Chicago, IL).

3 A pumpkin MYB1R1 transcription factor, CmMYB1, increased susceptibility to biotic stresses in transgenic tobacco

Powdery mildew (PM) is one of the major fungal diseases and can cause severe damage to pumpkin (*Cucurbita moschata*) crops worldwide. The causative agent of pumpkin PM is

第2章 南瓜抗白粉病相关基因的分离及功能研究

Podosphaera xanthii, a biotrophic fungus. To identify PM-related genes in pumpkin seedlings, cDNA representational difference analysis was previously performed using a RNA-seq method. One of the genes cloned from the transcriptome is homologous to pumpkin MYBR1-like gene encoding the R1-type MYB-like transcription factor. Here, we characterized this MYBR1-like homolog (named *CmMYB1*) from pumpkin and investigated its role in biotic stresses. The expression of *CmMYB1* in resistant pumpkin seedlings was dramatically induced by PM spores inoculation at the early infection stage (before 48 h), and was suppressed by exogenously sprayed hydrogen peroxide, hormones treatments (salicylic acid, abscisic acid, methyl jasmonate, ethephon), and salt stress using quantitative RT-PCR. Constitutive overexpression of *CmMYB1* in tobacco slightly aggravated the visible mildew symptoms of leaves, H_2O_2 accumulation and cell necrosis in seedlings inoculated with PM pathogens. Furthermore, the expression of *NPR1* and *PDF1.2* in the transgenic plants was lower than that in the wild-type plants. On the other hand, following inoculations with bacterial wilt and scab pathogens, the concentrations of bacterial pathogens near the infection sites were higher in the transgenic lines than in the wild-type plants, respectively. In conclusions, overexpression of *CmMYB1* in tobacco displayed increased susceptibility to PM, bacterial wilt and scab. *CmMYB1* transgenic tobacco may reduce stress resistance by downregulating hormone-mediated defense genes to aggravate the leaf damage caused by PM. *CmMYB1* may play negative roles in biotic stresses in tobacco.

3.1 Introduction

Pumpkin (*Cucurbita moschata* Duch.) is an important

vegetable crop cultivated worldwide. In China alone, 8,427,676 tons of pumpkin, squash and gourds (i.e., about 26.91% of the global yield) is harvested annually from 453,104 ha of land (i.e., 22.74% of the global area). Powdery mildew (PM), a biotrophic plant pathogen mainly caused by *Podosphaera xanthii*, is one of the major fungal diseases that can cause severe damage to Cucurbitaceae crops (Fukino et al., 2013; Pérez-García et al., 2009). The excessive application of fungicides is ineffective for controlling PM, because the associated increased selection pressure enables PM pathogens to develop increasing levels of fungicide resistance. Therefore, it is an effective way to study the resistance mechanism of powdery mildew in pumpkin and breed resistant varieties.

The MYB transcription factors (TFs) are one of the largest members of TF families. Each MYB DNA-binding domain located at N-terminus comprises 1 - 4 serial and non-redundant imperfect repeats (R1, R2, R3 and R4). Plant MYB proteins have been classified into three main groups: R2R3-MYB, R1R2R3-MYB, and a heterogeneous MYB-like group, which usually contain a single MYB repeat (Yanhui et al., 2006; He et al., 2016). The well-studied R2R3-MYB TFs are involved in various defense and abiotic stress responses (Hamama et al., 2017; Chen et al., 2015; Zhang et al., 2019). However, there are few functional studies on single MYB-like domain TFs in plants. On the basis of phylogenetic relationships (Yanhui et al., 2006) and classification schemes (Riechmann et al., 2000), the MYB-like proteins can be divided into five subgroups: CCA1-like, CPC-like, TBP-like, I-box-binding-like and R-R-type. A few studies revealed the diversity in MYB-like TF functions. For example, the putative R1-type MYB-like TF (*StMYB1R-1*) is responsive to drought stress and is unaffected by biotic stresses. The overexpression of *StMYB1R-1* increases the tolerance of potato plants to drought stress (Shin et al., 2011). The single MYB-like domain TFs (CCA1-like) are involved in the regulation of circadian rhythms (Alabadi et al., 2001). An *Arabidopsis* R-R-type MYB TF (*AtDIV2*) adversely affects salt stress responses by modulating abscisic acid (ABA) signaling (Fang et al., 2018). In rice, a putative R-R-type MYB TF (*MID1*) is involved in drought stress response; the overexpression of *MID1* increases the yield of rice under drought stress (Guo et al., 2016). Overexpression of *RsMYB1* in transgenic petunia can increase heavy metal stress tolerance by enhancing metal detoxification and antioxidant activities (Ai et al., 2018). The above-mentioned MYB-like TF functions were mainly reported to be involved in the response to abiotic stress in plants. However, MYB-like protein functions related to biotic stress resistance remain poorly characterized in plants.

To date, the molecular mechanism of pumpkin resistance to powdery mildew remains poorly unknown. Previously, we identified highly PM-resistant pumpkin materials (Zhou et al., 2010) and generated transcriptomes from the leaves of the highly-resistant pumpkin inbred line after inoculation with PM pathogens (Guo et al., 2018). An earlier analysis of PM-infected pumpkin identified 4 716 differentially expressed genes. In this study, we isolated and characterized the function of a MYBR1-like homolog (*CmMYB1*) in pumpkin from the transcriptomes. Overexpression of *CmMYB1* in tobacco leads to increased susceptibility to PM, bacterial wilt and scab.

3.2 Materials and Methods

3.2.1 Plant material and treatments

Pumpkin inbred line 112-2 and cv. 'Jiujiangjiaoding' (abbreviated

'JJJD'), highly resistant and susceptible to PM respectively, were used in this study (Zhou et al., 2010). Seedlings were grown in 1 ∶ 1 mixture of soil and peat and placed in a growth room with day/night temperatures of 28/18℃ under light/dark photoperiods of 15 h and 9 h. Seedlings at the third-leaf stage were used to establish the following treatments. The whole plants were sprayed with a freshly prepared spore suspension (10^6 spores/mL), exogenous 1.5 mmol/L hydrogen peroxide (H_2O_2) (Phygene, China), 100 μmol/L salicylic acid (SA) (Pelemix, China), 100 μmol/L ABA (Pelemix, China), 100 μmol/L methyl jasmonate (MeJA) (Pelemix, China), 0.5 g/L ethephon (ETH) (Pelemix, China) and 0.4 mmol/L NaCl (GuoYao, China), respectively (Guo et al., 2019). *P. xanthii* spores, which collected from naturally infected pumpkin leaves in a local filed, were quickly brushed into sterile water with a soft brush and diluted at a concentration 10^6 spores/mL under a microscope. Moreover, water alone was used for the control treatment (CK) under identical controlled conditions. Two young leaves from four separate seedlings were collected after 0, 3, 6, 12, 24, and 48 h to form one sample and frozen in liquid nitrogen for further RNA extraction to examine *CmMYB1* expression patterns in response to PM and exogenous treatments.

3.2.2 Isolation of *CmMYB1* cDNA and sequence analysis.

The MYBR1-like homolog EST (GenBank No. SRR5369792) was obtained from a transcriptome constructed with highly-resistant pumpkin seedling inoculated with *P. xanthii* pathogens (Guo et al., 2018). The full-length open reading frame (ORF) of the MYBR1-like homolog was cloned using a homology-based candidate gene method (Zhu et al., 2011). The protein sequence was characterized using ExPASy tool for theoretical molecular weight (Mw) and isoelectric point (pi) (jellqvist et al., 1993) and ClustalW for conserved MYB domain prediction (Thompson et al., 1994). The phylogenetic tree was constructed using Mega 5.0 by a neighbor-joining method (Guo et al., 2015).

3.2.3 Generation of CmMYB1 transgenic tobacco plants

Genetic transformation of pumpkin remains refractory, and there are only three reports of successful transformation using the combination of vacuum infiltration and Agrobacterium infection (Nanasato et al., 2014; Nanasato et al., 2011; Chen et al., 2021). The full-length sequence of *CmMYB1* was amplified with forward and reverse primers with an added *Bam*H I and *Kpn* I sites respectively, and then inserted into the pVBG2307 vector under the control of the 35S promoter. The recombinant plasmid was introduced into *Agrobacterium tumefaciens* GV3101 using electroporation. Transgenic tobacco (*Nicotiana tabacum* cv. NC89) plants were obtained through Agrobacterium-mediated transformation according to the method described previously (Li et al., 2012). The transgenic T0 seeds were sown on the 1/2 MS medium containing kanamycin (Km). Next, the Km-resistant seedlings were examined by PCR analysis of a report gene (*NPTII*) and *CmMYB1*. Homozygous transgenic T2 lines (L1 and L3), which produced 100% Km-resistant plants in the T3 lines, were selected for the next analysis.

3.2.4 Primer design

All corresponding primers used in this study are listed in Table 2-5.

Table 2-5 Primers used in this investigation

Gene	Accession	Primer sequence (5'-3')
CmMYB1	MH105821	F: GAGAAAGAGAGAGAAAATGGTGAGG
		R: ACCGCAACTAAACTAAGATGCTGTC
RT-qPCR for *Cm MYB1*		F: CTCTCAGATGACCTCATTACCA
		R: AGTCGGAGTTCTCGTAGTAACA

(续)

Gene	Accession	Primer sequence (5'-3')
Over-expression vector		F: GGG<u>GGATCC</u>GAGAAAGAGAGAGAAAATGGTGAGG (*BamH* I) R: GGG<u>GGTACC</u>ACCGCAACTAAACTAAGATGCTGTC (*Kpn* I)
NtNPR1	U76707	F: ACATCAGCGGAAGCAGTAG R: GTCGGCGAAGTAGTCAAAC
NtPR1a		F: CCTCGTACATTCTCATGGTCAAT R: CCATTGTTACACTGAACCCTAGC
NtPR5		F: CCGAGGTAATTGTGAGACTGGAG R: CCTGATTGGGTTGATTAAGTGCA
NtPDF1.2	T04323	F: GGAAATGGCAAACTCCATGCG R: ATCCTTCGGTCAGACAAACG
NtPAL	X95342	F: GTTATGCTCTTAGAACGTCGCCC R: CCGTGTAATGCCTTGTTTCTTGA
NtEF1-α	AY206004	F: TGTGATGTTTTTGTTCGGTCTTTAA R: TCAAAAGAAAATGCAGACAGACTCA
β-actin		F: TCTCTATGCCAGTGGTCGTA R: CCTCAGGACAACGGAATC
NPTII		F: AGACAATCGGCTGCTCTGAT R: TCATTTCGAACCCCAGAGTC

Note: F: forward primer, R: reverse primer; 5'-3': N terminal-C terminal of cNDA sequence.

3.2.5 Performance of transgenic lines inoculated with Golovinomyces cichoracearum pathogens

The transgenic tobacco seedlings at the fifth-leaf stage were sprayed with a freshly prepared spore suspension (10^6 spores/mL). Tobacco PM pathogens, belonging to *Golovinomyces cichoracearum*, were collected from naturally infected *Nicotiana tabacum* leaves at Guizhou academy of tobacco sciences of China. The upper second leaves from the transgenic and wild type (WT) seedlings continued to grow in vitro for 10 days post-inoculation (dpi) (Guo et al., 2019) and were used to calculate the disease index as described previously (Ishii et al., 2001). Additionally, these upper second leaves of the transgenic and WT seedlings were harvested symmetrically along the sides of the main vein after 0, 3, 5, and 7 dpi to examine PM pathogens growth, H_2O_2 accumulation and cell death, respectively. The pathogens growth was stained with methyl blue (Guo et al., 2018), H_2O_2 accumulation and cell death were stained with 3, 3'-diaminobenzidine (DAB) and trypan blue as described previously (Choi et al., 2012), respectively. These upper second leaves were sampled at 0, 12, 24, 48, 72, and 120 h post-inoculation (hpi) for a extraction of total RNA to examine the expression of five hormone-related marker genes (*NtNPR1*, *NtPR1a*, *NtPR5*, *NtPDF1.2* and *NtPAL*).

The bacterial solutions (10^8 CFU/mL) of bacterial wilt pathogens and scab pathogens, which belong to necrotrophic *Ralstonia solanacearum* and hemi-biotrophic *Xanthomonas euvesicatoria* respectively, were infected into the lower epidermis of leaves from transgenic plants with a needleless syringe, respectively, and the concentration of the bacteria near the injection sites from these plants was determined after 6 dpi as described previously (Guo et al., 2019). At least six seedlings were contained in each sample, and the treatments were arranged in a randomized complete block design with three replicates.

3.2.6 qRT-PCR analysis

Leaves of each sample were used for total RNA isolation using the TRIZOL reagent (Invitrogen, USA). The PrimeScript™ first-strand cDNA Synthesis Kit was used for cDNA synthesis. Gene expression analyses were performed with qRT-PCR in triplicate on 96-well plates using SYBR© Premix Ex Taq™ Ⅱ (TaKaRa, Japan). The relative expression levels were calculated by the $2^{-\triangle\triangle CT}$ comparative method. The *β-actin* (Wu et al., 2010) and *NtEF1-α* (Yan et al., 2014) genes were used as an internal control for normalization of variations in samples of pumpkin plants and tobacco plants, respectively.

3.2.7 Statistical analysis

Valueswere expressed as the mean ± standard error of three independent replicates. Data were analyzed using one-way analysis of variance. post hoc comparison tests were performed with SPSS 19.0 software (SPSS Inc, Chicago, IL) to differ entate the means between treatments and control plants. The asterisk indicates a significant difference at $P < 0.05$.

3.3 Results

3.3.1 Isolation and Analysis of CmMYB1

The full-length sequence of this MYBR1-like homolog (named *CmMYB1*) was submitted to the GenBank database (accession number MH105821). The gene comprised 892 bp, including an 855 bp ORF encoding 285 amino acids. The predicted polypeptide had a pI of 9.9 and a Mw of 31.6 ku. An alignment of the deduced CmMYB1 amino acid sequence with homologous sequences is presented in Supplementary Fig. S1. The predicted amino acid sequence included a conserved MYB domain (amino acids 95-145) and a conserved domain search analysis of CmMYB1 identified a SANT domain with a conserved R1 MYB repeat. In the phylogenetic tree conducted on the basis of the amino acid sequences of CmMYB1 and other known MYB proteins (Supplementary Fig. S2), CmMYB1 was clustered with its homologues from *C. pepo*, *Cucumis melo* and *Cucumis sativus*.

3.3.2 CmMYB1 expression patterns in response to PM and exogenous treatments

The *CmMYB1* expression patterns in both the resistant inbred line 112-2 and susceptible cv. 'JJJD' were analyzed following a PM infection and exogenous treatments (Fig. 2-16). In response to PM, the *CmMYB1* expression in 112-2 plants was significantly upregulated at 3, 6, 24, and 48 hpi, but downregulated at 12 and 72 hpi. The MeJA and ETH treatments significantly downregulated *CmMYB1* expression in 112-2 plants, except at 12 and 24 hpi, respectively. The opposite expression of *CmMYB1* appeared in the susceptible 'JJJD' under PM, MeJA, and ETH treatments. Following the H_2O_2, SA, ABA, and NaCl treatments, *CmMYB1* expression was significantly lower in both two pumpkin seedlings than in the control seedlings, except for the 112-2 seedlings at 24 h after the SA treatment and 12 h after the NaCl treatment. More specifically, the *CmMYB1* expression level in 112-2 plants was downregulated more than 10 fold by H_2O_2 and SA treatments during the study period.

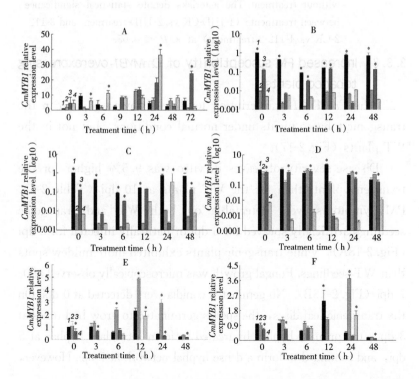

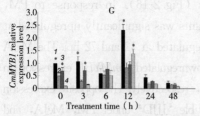

Fig. 2-16 Expression profiles of *CmMYB1* in pumpkin in response to powdery mildew and exogenous treatments. The pumpkin seedlings were sprayed with 10^6 spores/mL spore suspension (A), 1.5 mmol/L exogenous H_2O_2 (B), 100 μmol/L SA (C), 100 μmol/L ABA (D), 100 μmol/L MeJA (E), 0.5 g/L ETH (F) and 0.4 mmol/L NaCl (G). Values represent the means ± SD based on three biological repetitions. Control-plants without treatment. The asterisks denote statistical significance between treatments (1-JJJD-CK vs. 2-JJJD-treatment and 3-112-2-CK vs. 4-112-2-treatment) at * $P < 0.05$.

3.3.3 Increased PM susceptibility of *CmMYB1*-overexpressing tobacco plants

The *CmMYB1* transcription level was obviously high in the transgenic tobacco plants under normal conditions, but not in the WT plants (Fig. 2-17).

Disease severity of leaves in vitro was 9.5% higher for the transgenic plants than for the WT plants at 10 dpi (Table 2-6). PM symptoms were detectable on both WT and transgenic seedlings artificially sprayed for 6 dpi or naturally infected for 8 dpi (Fig. 2-18A), while transgenic plants exhibited more mildew spots than WT seedlings. Fungal growth was microscopically observed at 0-7 dpi (Fig. 2-18B). No geminated conidia were detected at 0 dpi. On the transgenic leaf discs, the spores germinated to grow bud tubes at 3 dpi, the primary hyphae bifurcated to form secondary hyphae at 5 dpi, and continued to form a dense hyphal network at 7 dpi. However,

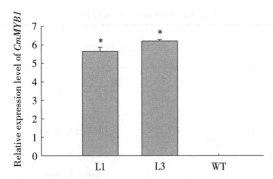

Fig. 2-17 Expression profiles of *CmMYB1* in tobacco plants under normal growth conditions. The error bars indicate the standard deviation from triplicate experiments. Asterisks indicate significant difference (* $P < 0.05$) compared with WT plants. WT, wild type; L1 and L3, transgenic lines.

a number of germinating spores at 3 dpi and secondary hyphae at 5 dpi were observed relatively more on the transgenic leaf discs than those detected on the WT leaf discs, respectively. Overall, the PM pathogens growth on the transgenic leaves was slightly faster than that on the WT leaves as determined by both the visual estimates and microscopic observation. Large brown spots on the transgenic leaves showed no obvious difference at 3 dpi. Larger and more intensely brown spots were observed on the transgenic leaves than these on the WT leaves at 5 and 7 dpi (Fig. 2-18C), indicating that *CmMYB1* overexpression promoted the accumulation of H_2O_2 in transgenic plants during the late PM infection stage. In addition, blue spots were initially detected on the transgenic leaves at 3 dpi and increased in size and number at 5 and 7 dpi compared with the WT leaves (Fig. 2-18D), implying that *CmMYB1* overexpression in the transgenic plants also accelerated cell death as the PM infection progressed.

Table 2-6 Disease severity of leaves of tobacco seedlings infected with powdery mildew

Materials	Disease severity (leaves in vitro)
L1	20.13±1.45 a
L3	19.29±1.32 a
WT	18.00±1.50 a

Data are mean values (± SD) of at least three independent experiments. Different lowercase letters indicate significantly different between treatments ($P < 0.05$). WT- wild type; L1 and L3 are two independent transgenic lines with *CmMYB1* overexpression.

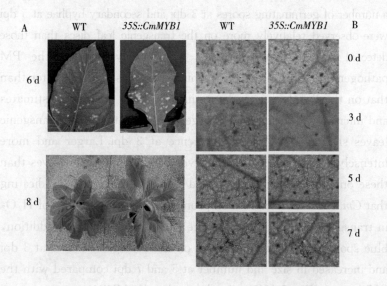

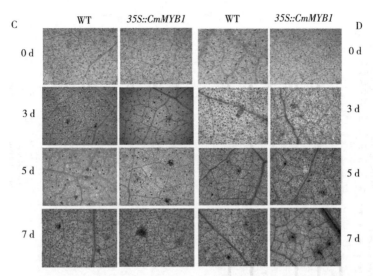

Fig. 2-18 Powdery mildew resistance conferred by *CmMYB1* expression in transgenic tobacco plants. The mildew symptoms of transgenic plants (35S∷*CmMYB1*) and wild type (WT) at 6 and 8 days post-innoculation with *Golovinomyces cichoracearum* (A); the fungal growth of leaves at different time points (B); scale bars, 100 μm; DAB staining (C) and trypan blue staining (D); scale bars, 200 μm.

3.3.4 Expression of signaling-Related genes in transgenic tobacco plants

To study the signal pathways regulated by CmMYB1 in the interactions between plant and PM pathogens, expression profiles of signaling-related genes in the SA, JA, and ethylene signal pathways were analyzed by qRT-PCR (Fig. 2-19).

In response to the PM pathogens infection, the *NPR1* (except at 12 hpi), and *PDF1.2* expression levels were lower in the transgenic plants than in the WT plants. Compared with WT plants, the *PR1a* expression level in the transgenic plants was lower at 12 and 24 hpi, but higher thereafter. Opposite *PAL*

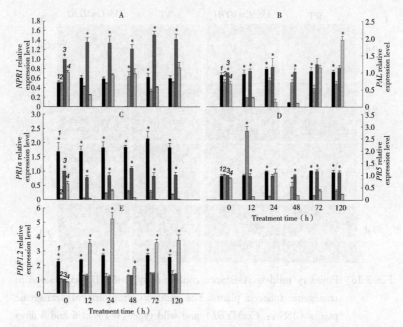

Fig. 2-19 Expression profiles of signal-related genes (A-E) in the wild-type and transgenic tobacco lines inoculated with powdery mildew. The samples of two T2 lines (L1, L3) were equally mixed to analyze genes expression by qRT-PCR. *NPR1* (A); *PAL* (B); *PR1a* (C); *PR5* (D); *PDF1.2* (E). 1-L1/3-CK-transgenic lines and 3-WT-CK-wild type under normal conditions; 2-L1/3-PM-transgenic lines and 4-WT-PM-wild type infected with powdery mildew. Asterisks denote significant difference between treatments and control (L1/L3-PM vs. L1/L3-CK and WT-PM vs. WT-CK) at * $P<0.05$.

expression patterns were observed higher before 48 hpi, but lower thereafter in the transgenic plants than in the WT plants. Furthermore, the *PR5* expression level was irregular (slightly higher in the transgenic plants than in the WT plants at 24 and 72 hpi, but lower at other time points). On the basis of these results, *CmMYB1* overexpression in transgenic plants decreased the transcription level of *NPR1* and *PDF1.2* in response to PM infection. The decreased PM resistance of the transgenic plants might be related to the downregulated expression of these genes.

3.3.5 Increased susceptibility of transgenic tobacco plants to bacterial diseases

To analyze the effect of CmMYB1 on other plant diseases, two common bacterial pathogens causing bacterial wilt and scab were injected into tobacco plants (Fig. 2-20). At 6 dpi, the severity of the chlorosis near the injection sites of the sixth leaf was

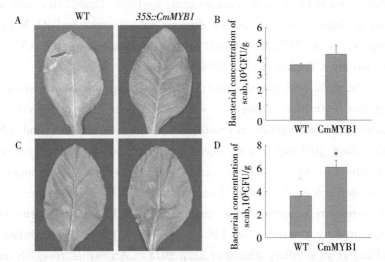

Fig. 2-20 The resistance of *CmMYB1* in tobacco plants to bacterial wilt and scab. Pathogens symptoms near the leaf injection sites of bacterial wilt bacteria (A); bacteria concentration near the leaf injection sites of bacterial wilt bacteria (B); pathogens symptoms near the leaf injection sites of scab bacteria(C); bacteria concentration near the leaf injection sites of scab bacteria (D). The error bars indicate the standard deviation from three biological repeats. Asterisks indicate significant differences (* $P<0.05$) compared with WT plants.

similar between the WT and transgenic plants. However, the bacterial wilt and scab pathogen concentrations were, respectively, 1.18 and 1.68 times higher in the transgenic plants than in the WT plants. Accordingly, *CmMYB1* overexpression slightly decreased the resistance of transgenic tobacco plants to bacterial wilt and scab.

3.4 Discussion

In this study, we isolated a novel pumpkin MYBR1-like gene *CmMYB1*. The predicted CmMYB1 amino acid sequence included a conserved MYB domain (amino acids 95-145). CmMYB1, which has a single conserved SANT domain with a conserved R1 MYB repeat, is a MYBR1-related protein belonging to the CCA1 subfamily (Yanhui et al., 2006).

Salicylic acid plays a central role in plant disease resistance against biotrophic pathogens, whereas JA is critical for activating plant defenses against necrotrophic pathogens. The SA and JA signalling pathways have been demonstrated to interact in a positive way to activate resistance to the two types of pathogens (Wasternack et al., 2013; Pieterse et al., 2012). The expression of stress responsive genes mediated by ABA and ethylene is mediated by a number of TFs, including MYB family members (Fang et al., 2018; Zhang et al., 2018). An earlier research on wheat seedlings proved that *TaMYB4* transcription in leaves was induced after treatments with different hormones (SA, ETH, ABA and MeJA), which might be related to the defense responses mediated by the crosstalk among the signaling pathways activated by SA, ETH and ABA (Al-Attala et al., 2014). The transcript levels of *VdMYB1* showed a significant increase after inoculation with the PM fungus, and no significant change under MeJA treatment (Yu et al., 2019). In the current study, the *CmMYB1* expression level in the resistant 112-2 plants was obviously repressed by H_2O_2, hormones (SA, ABA, MeJA, ETH), and NaCl treatments and was induced after inoculation with PM pathogens for 48 h. Additionally, overexpression of *CmMYB1* in tobacco plants increased the disease severity, accelerated H_2O_2 accumulation and cell necrosis as the PM infection progressed, suggesting that the overexpression of *CmMYB1* resulted in plant rapid H_2O_2 accumulation and HR-related necrosis to insufficiently prevent pathogen late invasion (i. e. disease severity was calculated at 10 dpi). Our findings are consistent with the results of an earlier investigation that proved *BnaMYB111L* in tobacco promotes reactive oxygen species production and hypersensitive response-like cell death (Yao et al., 2020). A R2R3-type MYB-related protein (MYB30) in *Arabidopsis thaliana* positively regulates programmed cell death associated with hypersensitive response (Marino et al., 2019). There is a different report that overexpression of *VdMYB1* resulted in plant rapid accumulation of ROS to prevent pathogen early invasion: ROS were strongly induced at 20 min after PM pathogens inoculation in transgenic leaves, however, with the treatment time extended to 60 min, showed no difference between overexpressing and WT plants (Yu et al., 2019). The effect of ROS accumulation on plant defense response may be different with the extension of pathogen infection time. In addition, the overexpression of *CmMYB1* herein also increases the susceptibility of tobacco to bacterial wilt and scab, which is consistent with the effects of *CmMYB1* overexpression on the susceptibility to PM. So, pumpkin *CmMYB1* gene may broadly play a negative role in plant response to diseases. Transgenic tobacco plants co-expression of *VdMYB1* and *VdSTS2* exhibit resistance to bacterial wilt (Yu et al., 2019).

The protein encoded by *PDF1.2* is important for the JA/ETH-dependent signaling pathway. The *NPR1*, *PAL*, *PR1a*, and *PR5* expression levels are markers for the SA signaling pathway. In an earlier investigation, the inoculation of *TaMYB* 4-knockdown plants with an avirulent CYR23 strain resulted in significant decrease in *PR1*, *PR2*, *PR5* and *TaPAL* transcription levels (Al-Attala et al., 2014). Overexpression of *VdMYB1* in grapevine leaves positively regulated defense responses by not altering the expression of genes in SA- and JA-dependent pathways(Yu et al., 2019). In the current study, *NPR1* and *PDF1.2* were expressed at lower levels in the transgenic plants than in the WT plants in response to PM, suggesting that CmMYB1 negatively regulates the SA- and JA/ETH-dependent defense pathway by downregulating *NPR1* and *PDF1.2* expression in tobacco, respectively. We propose that the downregulated expression of *NPR1* and *PDF1.2* may be responsible for the compromised resistance to PM pathogens.

In conclusion, the results of this study indicate that overexpression of the pumpkin *CmMYB1* gene in tobacco leads to increased susceptibility to PM, bacterial wilt, and scab. Additionally, the increased susceptibility of *CmMYB1* overexpression may not be related to H_2O_2 accumulation and cell necrosis, but may be the downregulation of SA/JA-mediated defense genes. Our findings may facilitate the molecular breeding of PM resistance in pumpkin.

4 A Pathogenesis-Related Protein 1 of *Cucurbita moschata* responds to powdery mildew infection

4.1 Introduction

V. *Cucurbita* powdery mildew (PM) is one of the most

destructive diseases that acutely diminishes the productivity and quality of pumpkin crops globally. The causal agent of PM in pumpkin is *Podosphaera xanthii*, a biotrophic fungus widely distributed worldwide (Perez-Garcia et al., 2009; Fukino et al., 2013). The use of pesticides control PM is associated with premature leaf senescence in pumpkin and the risk of developing drug-resistant pathogens due to prolonged usage. Also, pesticides cause environmental pollution, which poses a health risk to animals and humans. Most cultivated pumpkin varieties are very susceptible to PM, especially when grown at high temperatures under dry and wet alternating conditions (Mccreight, 2003). Therefore, there is a need to screen for PM-resistant genes to accelerate the development of new pumpkin varieties with PM resistance.

PR1 (pathogenesis-related protein 1) proteins play a crucial role in plant defense responses. For example, PR1 accounted for about 2% of the total protein in tobacco leaves infected by pathogens (Alexander et al., 1993). PR1 proteins are grouped under a multigene superfamily known as CAP (cysteine-rich secretory protein, antigen 5, and pathogenesis-related-1 protein) (Gibbs et al., 2008) and can be broadly categorized as acidic or alkaline based on their theoretical isoelectric points. Most PR1 proteins contain stress signaling peptides such as CAPE-1 (CAP-derived peptide 1), which comprise the last 11 amino acids from the C-terminus of the PR1 protein (Chen et al., 2014). Plant *PR1* is typically considered an indicator of salicylic acid (SA) inducible systemic acquired resistance (SAR), a plant immune response that prevents further spread of infection to non-infected parts of the host plant. SAR also plays a crucial role in hypersensitive response (HR)-related cell death (Van et al., 2006). Recent reports have emerged demonstrating that *PR1* genes in various plants are involved in the response to many phytopathogens attack (Han et al., 2023; Soliman et al., 2019), especially fungi (Anisimova et al., 2021; Kiba et al., 2007). PR1 proteins are involved in sterol-binding activity of caveolin-binding motif (CBM) in the CAP region, which targets and inhibits phytopathogens in response to host infection (Wang et al., 2022; Gamir et al., 2017; Breen et al., 2017). PR1 proteins also facilitate cell wall thickening to limit the invasion and spread of pathogens in the apoplast (Wang et al., 2013). PR proteins have been extensively applied in the development of transgenic plants with broad-spectrum resistance to a wide range of pathogens. However, there are only a few studies on the function of PR1 proteins in *C. moschata* or related Cucurbita species during biotic stress responses.

In an earlier study, we identified pumpkin genotypes with high powdery mildew resistance through natural and artificial inoculations under field conditions (Zhou et al., 2010). Moreover, transcriptome sequencing revealed variations in the expression levels of many *PR* genes in pumpkin in response to *P. xanthii* infection (Guo et al., 2018). To clarify the role of PR proteins in response to *P. xanthii* infection, we isolated a PR1-like gene from pumpkin and named it *CmPR1* (*GenBank accession no. MH105818*). Additionally, the *CmPR1* expression profile and the subcellular localization of the encoded protein were analyzed. Finally, we overexpressed the *CmPR1* gene in tobacco to validate its function in disease resistance.

4.2 Materials and methods

4.2.1 Plant material and stress treatments

Seeds of two pumpkin genotypes inbred line 112-2 and cultivar

'JJJD' were provided by Henan Institute of Science and Technology, Henan Province, China. The inbred line 112-2 is highly resistant to *P. xanthii*, whereas cultivar 'JJJD' is susceptible. The pumpkin seeds were sown in plastic pots (9 cm deep) containing a 3∶1 mixture of grass charcoal and perlite and the resulting seedlings were grown as previously described by Guo et al. (2018). The seedlings at the fourth leaf stage (at approximately 4 weeks) were treated as previously described (Guo et al., 2019). In detail, the seedlings were sprayed with freshly-prepared spore suspension (10^6 spores/mL) and solutions containing 1.5 mmol/L H_2O_2, 100 mmol/L SA, 100 mmol/L abscisic acid (ABA), 100 mmol/L methyl jasmonate (MeJA), 0.5 g/L ethephon (ETH), or 0.4 mmol/L sodium chloride (NaCl) in separate treatments. Distilled water alone was used as the control treatment. Specially, SA was dissolved in 10% ethanol, and MeJA, ETH in sterile water. Control plants were sprayed with 10% ethanol and sterile water, individually (Tang et al., 2017). The treated seedlings were maintained in a growth chamber with a photoperiod of 15 h/9 h light/dark (28℃/18℃, 5 500 lux light intensity). Subsequently, two or three upper leaves of four separate seedlings were sampled to detect *CmPR1* expression profiles at 0, 3, 6, 12, 24, and 48 h time points under various stress conditions. The experiments were arranged in a completely random block design with three biological replicates.

4.2.2 *CmPR1* cloning and sequence analysis

The *PR1* EST (GenBank accession No. SRR5369792) was isolated from a transcriptome of pumpkin seedlings inoculated with *P. xanthii* pathogens (Guo et al., 2018). Full-length open reading frame (ORF) of *PR1* was cloned using cDNA sequence as a probe by a homology-based candidate gene method. The ExPASy

Compute pI/Mw tools was used to calculate the theoretical molecular weight (Mw) and isoelectric point (pI) of the CmPR1 protein. DNAMAN tool (version 6.0.40) was used to analyze sequence alignment of amino acids between pumpkin CmPR1 and other homologous PR1 proteins. Blast tool of the NCBI databases was used to search for homologous protein sequences (http://www.ncbi.nlm.nih.gov/blast). SignalP5.0 server was utilized to predict potential signal peptide regions and the cleavage sites.

4.2.3 Subcellular localization analysis

The *CmPR1* ORF was inserted into the pBI221-GFP vector to form a green fluorescent protein (GFP) -labeled CmPR1 fusion construst. Transient expression of the *CmPR1-GFP* fusion gene in *Arabidopsis thaliana* protoplasts was analyzed using polyethylene glycol method described by Lee et al. (2013). GFP fluorescence was directly imaged using a confocal fluorescence microscope (UltraVIEWVoX, Olympus, Japan) under the excitation wavelength of 488 nm and the capture wavelength of 448-508 nm. Cell membrane and nucleus were detected by 1, 10-dioctadecyl-3, 3, 30, 30 -tetramethylindocarbocyanineperchlorate (DiI) and 2- (4-Amidinophenyl) -6-indolecarbamidine dihydrochloride (DAPI) staining, respectively.

4.2.4 Plasmid vector construction and genetic transformation of tobacco

Full-length forward and reverse primers of *CmPR1* gene containing *Bam*HI and *Kpn* I sites, respectively, were used to amplify the cDNA fragment. The cDNA fragment was inserted into the cloning site of the pMD19-T vector (Takara, Japan) and then digested from the recombinant pMD19-T vector using restriction endonuclease *Bam*HI and *Kpn* I. The *CmPR1* cDNA fragment with the restriction enzyme sites was ligated to the *Bam*HI-*Kpn* I site of the expression vector pVBG2307 harboring 35S cauliflower mosaic virus promoter. The recombinant vector pVBG2307-*CmPR1* was validated by sequencing and double enzyme digestion and then transformed into *Agrobacterium tumefaciens* GV3101 by electroporation. Putative transgenic tobacco (*Nicotiana tabacum* L. cv. NC89) plants overexpressing *CmPR1* were generated via *Agrobacterium*-mediated transformation using the leaf disk method (Li et al., 2012). Subsequently, transgenic plants were identified by the kanamycin selective medium followed by PCR amplification of reporter gene *NPTII* and target gene *CmPR1*. More than three homozygous T2 transgenic lines were used for subsequent experiments.

4.2.5 Evaluation of disease resistance of transgenic tobacco overexpressing *CmPR1*

Transgenic tobacco plants overexpressing *CmPR1* were used to analyze resistance to powdery mildew at the six-leaf stage. The causative agent of tobacco powdery mildew (*Erysiphe cichoracearum* DC) was collected from naturally infected tobacco leaves. Spore suspension (10^6 spores/mL) of the pathogen was sprayed onto asymptomatic plants. Subsequently, various parameters were assessed, including disease incidence, growth of the PM fungus, H_2O_2 accumulation and cell death. The inoculated leaves in vitro were examined for the disease symptoms of powdery mildew (Guo et al., 2019). After PM inoculation, the petioles were wrapped with absorbent cotton wool and put into plastic box with a lid to moisturize. The inoculated plants were cultured at 28℃ in light/dark (16 h/8 h). Attention was paid to note the incidences of moisture addition during the period.

The infected leaves of transgenic plants andwild type (WT) were symmetrically excised along the sides of the main vein at different corresponding time points after pathogen inoculation. Subsequently, histological observations of fungal growth and host responses were performed as described by Guo et al. (2018). The growth of PM fungus and cell death around the infection sites were examined by methyl blue staining and trypan blue staining, respectively. H_2O_2 accumulation was detected by 3, 3-Diaminobenzidine (DAB) staining (Choi et al., 2012). At least 20 infection sites were examined in each of four randomly selected leaf parts in each experiment.

4.2.6 Gene expression analysis

Total RNA was extracted from pumpkin and tobacco leaves using the Trizol method (Invitrogen, USA) according to the manufacturer's instructions. Total RNA concentration and purity (OD260/OD280 and OD260/OD230 ratio) were measured to detect the quality. RNA degradation was examined using 1% agarose gel electrophoresis. The first strand of cDNA was synthesized from total RNA as a template using PrimeScript™ II 1st Strand cDNA Synthesis Kit (TaKaRa, Japan). Subsequently, the cDNA was used as a template for qRT-PCR, which was performed on *Cycler* iQ™ Multicolor PCR Detection System (Bio-Rad, USA) using TB Green™ Premix Ex Taq™ II (Tli RNaseH Plus) (Invitrogen, USA). Relative gene expression levels were calculated using the $2^{-\Delta\Delta CT}$ method. *CmPR1* gene expression was analyzed to test its response to pathogen infection and exogenous signal molecules. In addition, the expression of marker genes (*NtNPR1*, *NtPR1a*, *NtPR5*, *NtPDF1.2*, and *NtPAL*) associated with SA or JA/ETH signal transduction was examined to evaluate the defense response mechanism of tobacco under powdery mildew infection. Pumpkin *β-actin* gene (Wu and Cao, 2010) and tobacco *NtEF1-α* gene (Xiang et al., 2017) were used as internal controls for normalization of gene expression. All primers used in this study are listed in Table 2-7.

Table 2-7　Primers designed in this study

Gene	Accession	Primer sequence (5'-3')
CmPR1	MH105818	F: TAGGCAGAAGCAAGCAACAATAA R: ACATAGGCGAGCGGCCCTACTAA
RT-qPCR for CmPR1		F: TAGGCAGAAGCAAGCAACAATAA R: CCCTGTAGATGGTTTCGTTGTCC
NtNPR1	U76707	F: ACATCAGCGGAAGCAGTAG R: GTCGGCGAAGTAGTCAAAC
NtPR1a		F: CCTCGTACATTCTCATGGTCAAT R: CCATTGTTACACTGAACCCTAGC
NtPR5		F: CCGAGGTAATTGTGAGACTGGAG R: CCTGATTGGGTTGATTAAGTGCA
NtPDF1.2	T04323	F: GGAAATGGCAAACTCCATGCG R: ATCCTTCGGTCAGACAAACG
NtPAL	X95342	F: GTTATGCTCTTAGAACGTCGCCC R: CCGTGTAATGCCTTGTTTCTTGA
NtEF1-α	AF120093	F: TGTGATGTTTTTGTTCGGTCTTTAA R: TCAAAAGAAAATGCAGACAGACTCA
β-actin		F: TCTCTATGCCAGTGGTCGTA R: CCTCAGGACAACGGAATC
NPTII		F: AGACAATCGGCTGCTCTGAT R: TCATTTCGAACCCCAGAGTC

4.2.7 Statistical analysis

Data values are expressed as mean ± standard error (SE). Differences between various treatments were analyzed using one-way analysis of variance (ANOVA) and means separated by LSD test (post-hoc) at a P-value $\leqslant$ 0.05. Data analysis was performed using IBM SPSS Statistics 25 software (SPSS Inc., USA).

4.3 Results

4.3.1 Isolation and analysis of *CmPR1* gene from pumpkin

CmPR1 was cloned based on the transcriptome of a highly resistant pumpkin genotype inoculated with powdery mildew using PR1 EST sequence as a probe. The full length of the *CmPR1* gene (GenBank: MH105818) was 658 bp, translating to 198 amino acids with a molecular weight of 22.9 ku and a pI of 6.88. Based on NCBI conserved domains database and SMART database, CmPR1 protein contained a CAP superfamily domain structure, a signal peptide, and SCP domains. Sequence alignment of amino acids between pumpkin CmPR1 and other plant genes showed that CmPR1 amino acids were highly identical to those of *C. maxima* (98.9% identical), *C. pepo* (96.5% identical), *Benincasa hispida* PR1 (91.2% identical), and *Cucumis sativus* PR1 (90.7% identical). In additon, those proteins contained highly conserved CAP domain sequences (amino acids 58-194), confirmed to belong to the CAP superfamily. A SignalP analysis revealed signal peptide regions (amino acids 1-21) of PR1 proteins at the N terminal in alignment. The predicted cleavage site is indicated by a circle (Fig. 2-21).

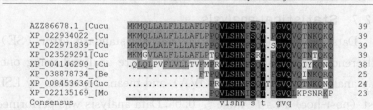

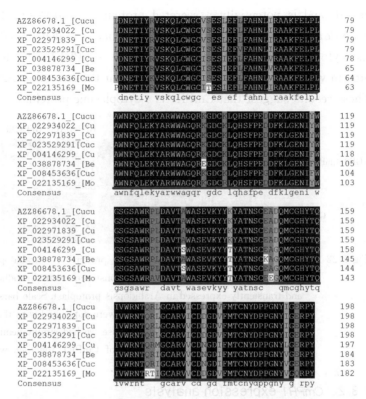

Fig. 2-21 Multiple sequences alignment of pumpkin CmPR1 amino acid with its homologs of different plant species. Origin of sequences: AZZ86678.1 [*Cucurbita moschata* cv. '112-2'], XP_022934022 [*Cucurbita moschata* cv. 'Rifu'], XP_022971839 [*Cucurbita maxima*], XP_023529291 [*Cucurbita pepo*], XP_004146299 [*Cucumis sativus*], XP_038878734 [*Benincasa hispida*], XP_008453636 [*Cucumis melo*], XP_022135169 [*Momordica charantia*]. The red colored underline indicates the signal peptide regions; the cleavage sites predicted are shown by the green circular. The blue underlines indicate the CAP domain structure of ~140 bp.

The deduced CmPR1 protein contains a predicted signal peptide and is potentially a secreted protein. The GFP signal in *Arabidopsis* protoplasts expressing GFP alone was distributed in the cytoplasm and nucleus. However, the CmPR1-GFP fusion protein was transiently expressed through out the cell (Fig. 2-22). These results suggest that CmPR1 is a cytoplasmic protein.

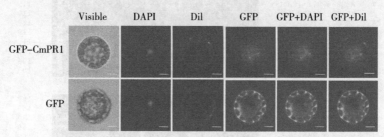

Fig. 2-22 Subcellular localization of CmPR1 protein in Arabidopsis protoplast. The fused GFP-CmPR1 constructs were introduced and transiently expressed in Arabidopsis protoplast. Scale bars = 5 μm. Note that images of GFP alone were duplicated with those described in our previous article because we had simultaneously expremented subcellular localization of CmPR1 together with CmSGT1 proteins (Guo et al., 2019).

4.3.2 *CmPR1* expression analysis

Compared withcontrol, *CmPR1* expression was significantly downregulated (0.13 fold) at 9 h, 12 h, and 24 h, but significantly up regulated (19.2 fold) at 48 h in resistant 112-2 seedlings inoculated with PM pathogens (Fig. 2-23), indicating that *CmPR1* was responsive to *P. xanthii* infection. Notably, H_2O_2 (Fig. 2-23B) and SA (Fig. 2-23C) treatments significantly inhibited *CmPR1* expression in control and inoculated seedlings. Specifically, *CmPR1* expression decreased by 0.02 fold in control under H_2O_2 treatment for 24 h and by about 0.05 fold in control under SA treatment, suggesting that *CmPR1* potentially plays a negative regulatory role in response to exogenous SA

第2章 南瓜抗白粉病相关基因的分离及功能研究

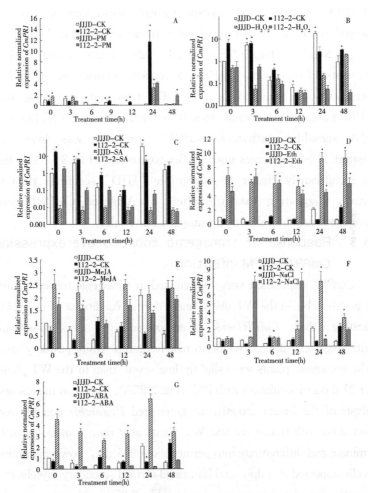

Fig. 2-23 Expression level of *CmPR1* gene in leaves of pumpkin plant after exogenous treatments. The seedlings were sprayed with a spore suspension of *Podosphaera xanthii* (A); exogenous hydrogen peroxide (H_2O_2) (B); salicylic acid (SA) (C); ethephon (ETH) (D); methyl jasmonate (MeJA) (E); NaCl (F); and abscisic acid (ABA) (G). The asterisks denote statistical significance between treatments (JJJD-CK vs. JJJD-treatment and 112-2-CK vs. 112-2-treatment) at $P < 0.05$.

and H_2O_2 treatments. An opposite *CmPR1* expression pattern was observed in the control and inoculated seedlings following ETH (Fig. 2-23D), MeJA (Fig. 2-23E), and salt stress (Fig. 2-23F) treatments. In particular, *CmPR1* expression increased by 16 fold in control at some time points after ETH treatment, indicating that *CmPR1* potentially responds to salt stressvia ethylene (ETH) and MeJA signaling pathways. *CmPR1* expression was lower in resistant 112-2 seedlings under ABA treatment (Fig. 2-19G), but was significantly higher in susceptible 'JJJD' seedlings than in control, suggesting that the effect of ABA on *CmPR1* gene expression may be related to pumpkin materials.

4.3.3 Response of transgenic tobacco overexpressing CmPR1 to PM infection

CmPR1 expression was up-regulated in transgenic tobacco plants but undetectable in the WT plants under normal conditions (Fig. 2-24), indicating that *CmPR1* was overexpressed in the transgenic plants. Regarding disease symptom manifestation, white powdery areas in the transgenic plants were slightly less severe than in the WT plants after 21 d post-inoculation with PM (Fig. 2-25A). Based on microscopic analysis of the fungal growth, no geminated *P. xanthii* conidia were detected on both transgenic and WT plants at 0 dpi, conidia began to germinate and differentiate into germtubes at 3.5 dpi, powdery mildew mycelia appeared at 5 dpi, and bifurcated formed secondary mycelia at 7 dpi in the inoculated transgenic leaves (Fig. 2-25B). However, in the inoculated WT leaves, conidia germinated and grew into mycelia at 3.5 dpi, dense hyphal network developed at 5 dpi, and chains of conidia formed at 7 dpi. These observations indicate that the growth and proliferation of the powdery mildew fungus were weaker in the transgenic plants than in the WT plants.

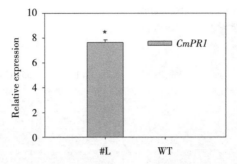

Fig. 2-24　The expression of *CmPR1* in tobacco plants under normal growth conditions. ♯L represents the equally-mixed samples of three independent *CmPR1*-overexpressed transgenic lines. asterisks denote statistical significance between both materials at $P<0.05$.

The DAB-stained brown spots in inoculated transgenic plants appeared at 2 dpi, expanded and deepened in color at 3 dpi, and continued to increase in continuous patches at 6 dpi (Fig. 2-25C). The DAB-stained brown spots were more in number and area in transgenic plants than in WT plants, and the differences were more obvious with the prolongation of infection time. In addition, the sporadic stained blue spots in the transgenic plants abounded at 4 dpi and continued to expand and increase between 5 and 7 dpi, becoming more and larger than in the WT plants (Fig. 2-25D). Overall, these results suggest that *CmPR1* overexpression in tobacco plants enhances reactive oxygen species accumulation and HR-cell death in response to powdery mildew.

4.3.4　Expression analysis of hormone signaling-related genes in tobacco

The *NPR1*, *PAL* (except at 0 h), and *PR5* (except at 120 h) expression level was higher in the PM-infected transgenic plants than in the CK plants, whereas these *NPR1*, and *PR5* (except at 24 h) genes level was lower in the PM-infected WT plants than in the CK

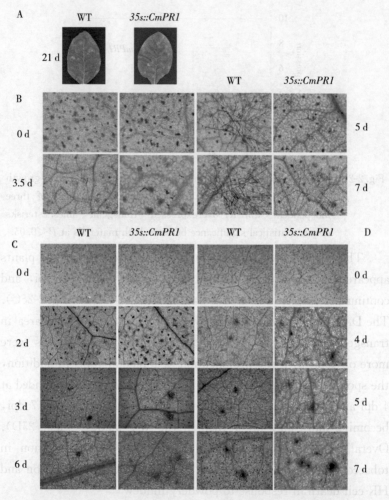

Fig. 2-25　Resistance analysis of tobacco leaves infected with *Erysiphe cichoracearum*. The pathogenic symptoms of transgenic tobacco (35s∷*CmPR1*) and wild type (WT) after 21d of infection, scale bars=2.0 cm (A); spores growth of the infected leaves (B); DAB-stained infected-leaves (C); Trypan blue-stained infected-leaves (D). Scale bars=200 mm.

plants. The *PDF1.2* expression levels were higher in the transgenic and WT plants infected with PM than in the CK plants, implying that PM induced the expression of this gene. The *PR1a* expression levels in the PM-infected WT plants were lower than in the CK plants, and the irregular pattern of this gene expression was observed in the transgenic plants infected with PM compared with CK plants. Under PM inoculation, the expression of *PDF1.2* and *PAL* (except 48 h) were wholly lower in the transgenic lines than in WT plants (Fig. 2-26). However, compared with WT-PM treatment, expression of *PR1a* in the PM inoculated transgenic lines was signifcantly up-regulated at 24 h, 72 h, and 120 h. Meanwhile, the expression of *PR5* in PM inoculated transgenic lines was significantly higher at 12 h and 48 h (24.7 and 4.1 fold of WT-PM), but slightly lower at 24 h and 120 h compared with WT-PM treatment. Notably, no significant difference in *NPR1* expression was observed between PM inoculated WT and transgenic lines. Altogether, these findings show that *CmPR1* overexpression in tobacco plants inhibits *PDF1.2* and *PAL* expression and induces *PR1a* expression in response to PM infection. Furthermore, the increased PM resistance of the transgenic plants appeared to be related to the upregulated expression of these genes.

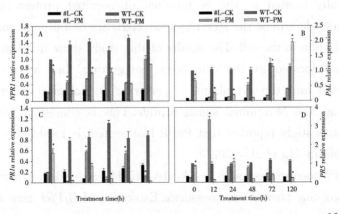

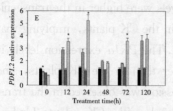

Fig. 2-26 Expression analysis of hormones signaling-related genes in tobacco infected with *Erysiphe cichoracearum*. The equally-mixed samples of three transgenic tobacco lines (#L) and wild type (WT) plants were used to analyze genes expression by qRT-PCR. A, *NPR1*; B, *PAL*; C, *PR1a*; D, *PR5*; E, *PDF1.2*. Asterisks denote significant difference between transgenic lines and WT plants (#L-PM vs. WT-PM and #L-CK vs. WT-CK) at $P<0.05$.

4.4 Discussion

In this study, a novel pumpkin PR1-like gene (*CmPR1*) was cloned. The predicted amino acid contained a conserved CAP domain and a signal peptide, indicating that CmPR1 is a putative pumpkin PR1 protein. CmPR1 protein is presumed to be acidic based on pI data. As a fragment of protein, a signal peptide is typically located at the N terminal of secreted protein. Signal peptides determine the secretory characteristics of a protein and its localization in the cell. The results of this study show that CmPR1 protein located in the cytoplasm, suggesting that it may be secreted into the cytoplasm to perform its biological functions through the N-terminal signal peptide. This is consistent with a previous study reported that PnPR-like protein is localized in the cytoplasm (Li et al., 2021).

Phytohormones, such as SA, JA, ETH, and ABA are involved in regulating plant disease resistance. Expression of *BjPR1* gene from

Brassica juncea was strongly induced by *Alternaria brassicae* infection and SA treatment, but not by JA or ABA treatments (Ali et al., 2018). *PnPR-1* transcripts from *Piper nigrum* were significantly up-regulated during *Phytophthora capsici* infection (Kattupalli et al., 2021). In this study, *CmPR1* expression was significantly up-regulated by late *Podosphaera xanthii* infection and exogenous application of ETH and MeJA, but significantly inhibited by SA and H_2O_2 treatments. This contradicts some previous studies that showed acidic PR proteins are up-regulated by various endogenous signaling molecules (ROS) and phytohormones (SA) in response to pathogen attack, while basic PR proteins are up-regulated by ETH and MeJA (Rigoyen et al., 2020). Our findings are consistent with some studies. For example, acidic *OgPR1a* expression level was strongly induced by exogenous ETH and JA (Shin et al., 2014). Although most of the PR1 proteins reported by previous studies were alkaline, it is not recommended to simply judge whether PR1 could enhance disease resistance based on pI value and sequence homology without conducting disease bioassays (Li et al., 2011).

The RNase activity of soybean *GmPRP* (*PR* gene) restricts the mycelial growth of *Phytophthora sojae* (Xu et al., 2014). Also, the RNase activity of jelly fig *PR-4* is related to its inhibitory effect on pathogens growth (Lu et al., 2012). In the present study, conidium germination and mycelium growth were significantly restricted in the transgenic tobacco plants expressing *CmPR1*, reflecting the role of *CmPR1* in enhancing disease resistance. However, further studies should be conducted to determine whether the inhibitory effect of CmPR1 protein on powdery mildew growth is related to RNase activity.

Plants initiate various defense responsesduring a pathogen attack. The first line of defense typically involves structural responses that include cell wall strengthening, and waxy epidermal cuticle development, and production of antimicrobial molecules (Hakim et al., 2018; Manghwar et al., 2022). However, many pathogens have evolved mechanisms to break the plant's first defense barrier; when this occurs, plants activate an alternative defense pathway, which is the metabolic modifications involving hypersensitive response, oxidative burst, synthesis of SA, ETH, and JA, and eventually the synthesis of PR proteins (Van et al., 2014; Das et al., 2011; Li et al., 2015). These defense responses are triggered based on the life style of the pathogen. For example, ROS generation and HR inhibit the growth of (hemi) biotrophs. On the contrary, necrotrophs stimulate ROS production to induce susceptibility-associated cell death of the host (Mengiste et al., 2012). In this study, DAB-staining revealed higher levels of H_2O_2 accumulation in the transgenic plants than in the WT plants. Excessive ROS accumulation induces cell death in plants (Petrov et al., 2012). Compared with inoculated WT plants, trypan blue staining showed that the transgenic plants exhibited more serious necrotic regions following the pathogen attack. This restricted the spread of the pathogen beyond the infection site, preventing further growth and spread to other plant parts. Overexpression of *CmPR1* in transgenic tobacco plants potentially limited the proliferation of powdery mildew pathogen by activating HR-related cell necrosis accompanied by ROS generation. In a similar study, TaTLP1 was shown to interact with TaPR1 to increase antifungal activity and inhibit fungal growth and cell death and H_2O_2 accumulation in *TaTLP1-TaPR1*-cosilenced plants were observed (Wang et al., 2020). Also, overexpression of *VpPR4-1* increased powdery mildew resistance of grape by repressing the growth of powdery mildew (Dai et al., 2016).

第 2 章 南瓜抗白粉病相关基因的分离及功能研究

Plant PR proteins facilitate defense responses to microbial pathogens via direct or indirect pathways. A direct response against invading microbial pathogen is typically characterized by inhibition of pathogen growth or spore germination. On the other hand, an indirect response involves PR isoforms, which play a more crucial role in plant resistance against pathogens. Plant pathogen invasion is quickly followed by activation of defense signaling pathways such as SA and JA, which induces the accumulation of PR proteins to minimize pathogen load or disease onset in uninfected plant organs. At intermediate SA levels, NPR1 (nonexpresser of PR1) accumulates and interacts with the TGA transcription factor, functioning as a coactivator of SA-responsive genes, including *PR* genes (Caarls et al., 2015). The *PAL*, *PR1a*, and *PR5* expression levels are markers of the SA signaling pathway. The *PDF1.2* gene is important for the JA/ETH-dependent signaling pathway. In this study, following a PM infection, the *PR1a* expression level was higher in the transgenic plants than in the WT plants, whereas the opposite pattern was observed for the *PAL* and *PDF1.2* expression levels. This suggests that in the SA pathway, the transactivation of *PR1a* is dependent on *CmPR1*, whereas the transactivation of *PAL* is unaffected by *CmPR1*. Additionally, *CmPR1* does not directly affect the JA/ETH-dependent defense pathway to regulate *PDF1.2* expression. Notably, *CmPR1* overexpression in tobacco alleviated the symptoms of tobacco powdery mildew by activating SA defense signaling pathways without inducing PAL expression level and suppressing the activities of JA/ETH-dependent defense pathways. This is consistent with previous studies that have revealed that the resistance to biotrophic and hemibiotrophic pathogens is often driven by SA signaling, while resistance to necrotrophic pathogens is mediated by JA and ETH signaling (Pieterse et al., 2012; Guo et al., 2019; Guo et al., 2020).

The results of this study indicate that heterologous expression of *CmPR1* gene from pumpkin enhances the resistance of transgenic tobacco plants to PM. Additionally, we revealed that overexpression of *CmPR1* gene in tobacco plants decreases the development of mildew symptoms caused by *E. cichoracearum*, possibly by inducing ROS accumulation and HR near infected sites, which further activates SA defense signaling pathway in uninfected leaves. This study provides new insights into understanding the defense mechanisms of Cucurbita crops against fungal diseases and can be exploited to develop disease-resistant crop varieties.

5 南瓜白粉病相关基因 *CmERF* 的克隆及功能分析

在植物中 AP2/ERF 类转录因子是转录因子家族中的一大类，根据 DNA 结合区域的数目 AP2/ERF 类转录因子分为两个亚家族：AP2 家族和 ERF 家族。AP2 家族成员在调控植物的生长发育方面发挥重要作用，ERF 家族主要参与植物对生物和非生物等胁迫的调控。*AtERF14* 在拟南芥中过量表达，上调下游防卫基因表达，并且 *AtERF1* 和 *AtERF2* 基因的表达依赖于 *AtERF14* 表达，而 *AtERF14* 基因缺失，突变体增加，对尖镰孢菌（*F. oxysporum*）引起的枯萎病敏感，说明 *AtERF14* 在拟南芥枯萎病抗性反应中发挥重要作用（Onate-Sanchez et al., 2007）。在柴胡细胞中，*BkERF1*、*BkERF2.1* 和 *BkERF2.2* 的瞬时表达可刺激防御相关基因 *PDF1.2* 和病程相关蛋白基因的表达，转基因拟南芥提高了对灰霉病（*B. cinerea*）的抗性，并诱导防御相关基因 *PDF1.2* 的表达（Liu et al., 2011）。在拟南芥中 *ERF1* 的超量表达提高了转基因植株对灰霉病（*Botrytis cinerea*）和番茄织

第2章 南瓜抗白粉病相关基因的分离及功能研究

球壳菌萎蔫病（*Plectosphaerella cucumerina*）的抗性，但对病原细菌丁香假单胞菌（*Pseudomonas syringae*）的抵抗力减弱（Berrocal-LoboM et al.，2002）。Tsi1 是一个 ERF 蛋白，能够与 GCC 盒和 CRT/DRE 元件结合，超量表达 *Tsi1*（Tobacco stress-inecued gene1）的烟草对病原菌的侵染表现出较高的耐性（Park et al.，2001）。植物 AP2 类转录因子可以提高植物对病原菌的抗性。*GmERF3* 基因在转基因烟草中过量表达可以诱导 *PR* 基因的表达，增强植物对青枯病（*Ralstonia solanacearum*）、烟草赤星病（*Alternaria alternata*）和 TMV 等病害的抗性（Zhang et al.，2009）。ERF 类转录因子可以正调控或负调控防御基因的表达。在正常的非胁迫条件下，转基因烟草中过量表达 *GmERF3* 基因可以诱导 *PR1*、*PR2*、*PR4*，*Osmotin* 和 *SAR8.2* 基因的表达。ERF 是 ET、JA、SA 等信号转导途径中的调控因子，家族成员参与调控这些相对独立的信号转导途径，也可被 ABA、ET 及 JA 诱导表达（Xu et al.，2008）。大白菜 *BrERF11* 和辣椒 *CaERF5* 的表达间接影响植株对青枯病的抗性，推测是由于该基因通过 SA、JA、ET 途径介导了抗病基因的表达（Lai et al.，2013）。烟草中 *NtERF5* 过量表达增强了烟草对烟草花叶病毒的抗病性（Fischer et al.，2004），番茄 *SlERF3* 的过量表达促进了 *PR* 基因（*PR1*、*PR2* 和 *PR5*）的表达，显著提高了对青枯病的抗性（Pan et al.，2010）；辣椒基因 *CaERF18* 的过量表达提高了辣椒对根结线虫的抗性。

5.1 材料与试剂

5.1.1 植物材料

南瓜：中国南瓜高抗白粉病自交系 112-2 和高感白粉病自交系'九江轿顶'，保存于河南科技学院南瓜团队。

烟草：烟草 NC89 为本实验室保存。

病原菌：南瓜白粉菌（*Sphaerotheca fuliginea*）、烟草白粉菌（*Erysiphe cichoracearum DC.*）、烟草青枯菌（*Ralstonia solanacearum* Yabuuhi)、烟草细菌性疮痂菌（*B. Xanthomonas*）。

5.1.2 菌株与质粒

菌株：大肠杆菌（*Escherichia coli.*）TOP10，购自天根生化科技有限公司；土壤农杆菌（*Agrobacterium tumefaciens*）GV3101，为本实验室保存。

质粒：克隆载体 pUC-T Vector 购自 TaKaRa 公司；超量表达载体 P1301 为本实验室保存。

5.1.3 试验试剂及仪器

PCR 用酶 TaKaRa LA Taq® 和 2×Es Taq MasterMix 以及限制性内切酶 *Xba*I、*Kpn*I 和 *Bam*HI，反转录试剂盒和实时定量试剂盒 TB Green™ Premix Ex Taq™ Ⅱ（Tli RNaseH Plus）(TakaRa) 均购自 TaKaRa 公司；琼脂糖胶回收试剂盒和质粒提取试剂盒购自天根生化科技有限公司。氨苄（Amp）、羧苄西林（Carb）、头孢霉素（Cef）、潮霉素购自 TaKaRa 公司。

本试验所用的主要仪器包括：高速冷冻离心机（Thermo）、紫外凝胶成像系统（Tocan 360）、超微量分光光度计（Thermo）、超低温冰箱（海尔 DW-86L486）、振荡培养箱、常规 PCR 仪（Biometra）、定量 PCR 仪（Bio-RAD）等。

5.2 方法与步骤

5.2.1 材料处理

(1) 南瓜白粉病接种处理及叶片采集　将 112-2 和'九江轿顶'播种于装有灭菌基质的穴盘中，南瓜幼苗 2 叶 1 心时，将田间自然发病叶片上的白粉菌，制成浓度为 $2.5×10^5$～$5×10^5$ 个/mL 的白粉菌分生孢子悬浮液，喷雾法接种南瓜幼苗，清水处理为对照。在温度 25℃/18℃（昼/夜）、相对湿度60%～70%、光周期 12 h、光照度为 4 400 lux 的光照培养箱中。分别于接种后 0、3、6、12、24、48 h 进行取样，用锡箔纸包裹，放入液氮中速冻，−80℃保存备用。

选取长势相同的 2 叶 1 心期 112-2 和'九江轿顶'南瓜幼苗，

第2章 南瓜抗白粉病相关基因的分离及功能研究

将 100 μmol/L SA、1.5 mmol/L H_2O_2、100 μmol/L ABA、0.4mmol/L NaCl、100 μmol/L MeJA（茉莉酸甲酯）和 0.5 g/L Eth（乙烯利）均匀喷洒至南瓜植株上，直至有水滴滑落，清水处理作为对照，分别在 0、3、6、12、24、48 h 采集叶片，每个处理每个时间点设置 3 个重复，液氮速冻，-80℃保存。

（2）烟草材料处理　烟草种子用 10% 次氯酸钠溶液消毒 10 min，其间不断摇晃，无菌水清洗 3 次，播种于无菌 MS 培养基上，4℃培养 3 d 后转移至组培实验室。生长 50 d 后用于转化烟草。

烟草叶片接种白粉病参照南瓜叶片白粉病接种方法。对幼苗期转基因植株通过喷雾法接种白粉菌，叶片正面喷洒均匀，在接种后 0、12、24、48、72、120 h 取样，保存于液氮中。对 5 叶 1 心期转基因烟草植株和未转基因野生型（对照植株）的真叶采用喷雾法接种白粉菌分生孢子悬浮液，叶片正面喷洒均匀，沿着主脉两侧对称取样，4 次重复，用于台盼蓝染色和 DAB 染色。

5.2.2　总 RNA 的提取及 cDNA 的合成

（1）总 RNA 的提取及检测　采用 Trizol 法提取南瓜及烟草总 RNA，具体方法参照试剂说明书。测取总 RNA 的浓度以及 OD_{260}/OD_{280}、OD_{260}/OD_{230} 比值，1% 琼脂糖凝胶电泳，检测 RNA 是否降解。

（2）cDNA 第一链合成　反转录以 RNA 为模板，参照 PrimeScript™ RT reagent Kit with gDNA Eraser 的操作说明书进行 cDNA 第一链合成。

5.2.3　实时定量 PCR

以 cDNA 第一链为模板，按照实时定量试剂盒 TB Green™ Premix Ex Taq™ II（Tli RNaseH Plus）的操作说明，在定量 PCR 仪上进行 qRT-PCR 扩增。实时定量引物如下（表2-8）。

表 2-8 实时定量特异引物序列

基因名称	引物序列（5′-3′）
CmERF	F：TGTCGTCCGATGGTGTCTA R：GACTGAAGGGTATCAAGGTAAT
NtNPR1	F：ACATCAGCGGAAGCAGTAG R：GTCGGCGAAGTAGTCAAAC
NtPR1a	F：CCTCGTACATTCTCATGGTCAAT R：CCATTGTTACACTGAACCCTAGC
NtPR5	F：CCGAGGTAATTGTGAGACTGGAG R：CCTGATTGGGTTGATTAAGTGCA
NtPDF1.2	F：GGAAATGGCAAACTCCATGCG R：ATCCTTCGGTCAGACAAACG
NtPAL	F：GTTATGCTCTTAGAACGTCGCCC R：CCGTGTAATGCCTTGTTTCTTGA
NtEF1-α	F：TGTGATGTTTTTGTTCGGTCTTTAA R：TCAAAAGAAAATGCAGACAGACTCA

注：F-正向引物；R-反向引物。

反应程序：95℃预变性 1 min，45 个循环；95℃ 变性 15 s，55℃ 退火 20 s，72℃延伸 30 s。在 55℃退火过程中采集荧光。

5.2.4 全长克隆及序列分析

（1）cDNA 合成　本试验操作步骤根据 TaKaRa 公司的 PrimeScript™ Ⅱ 1st Strand cDNA Synthesis Kit 的操作进行。

（2）目的基因的扩增（表 2-9）　根据 TaKaRa LA Taq® 说明书，引物序列和反应体系如下：

表 2-9 全长扩增特异引物序列

基因	编号	引物序列（5′-3′）	扩增大小（bp）
CmERF	MH105819	F：CTTTCCTTGGTGAAAAATGGTGAAG R：ATCAGATTGTGGAGCTTCTGCTGA	697

注：F 为正向引物；R 为反向引物。

（3）胶回收　参照胶回收试剂盒说明书进行操作。

(4) 连接 将 0.2 mL 离心管插到冰中预冷，按照下列体系（表 2-10）加入以下反应液：

表 2-10 连接反应体系

反应组分	体积（μL）
pUC-T	1.0
DNA	3.0
2×快速连接反应缓冲液	5.0
快速 T4 DNA 连接酶	1.0
总体积	10.0

轻轻混匀后，瞬时离心；25℃，连接 5 min，连接产物－20℃保存或直接用于转化。

(5) 转化连接产物 参照大肠杆菌 TOP10 试剂盒说明书。

(6) 重组质粒的 PCR 检测与测序

①在超净工作台中用无菌牙签挑取 5 个白色单菌落，置于含有 50 mg/L Amp 的 20 mL 液体 LB 培养基中，在 37℃震荡培养箱中培养过夜。

②在超净工作台中取 10 μL 活化后的菌液备用。

③菌液 PCR 检测。

④用无菌的 50%甘油保存菌液，于上海生工生物技术有限公司进行测序。

(7) 抗白粉病相关基因的生物信息学分析 用 DNAstar 软件对序列进行 ORF（Open Reading Frame）查询并预测氨基酸序列；利用 SMART 软件（http://smart.embl-heidelberg.de/）进行保守性结构域分析；利用 NCBI 数据库中 BLAST（http://blast.ncbi.nlm.nih.gov/Blast.cgi）进行同源蛋白基因查寻；利用 DNAMAN 软件进行同源蛋白序列的比对。

5.2.5 烟草遗传转化

(1) 构建超量表达载体 在目的基因 ORF 的两端设计含有

BamHI 和 KpnI 的酶切位点的引物（表 2-11），扩增基因序列。目的片段胶回收、连接 pUC-T、大肠杆菌感受态转化、质粒提取和测序，用限制性内切酶 BamHI 和 KpnI 酶切经过测序的质粒和 P1301 超量表达载体，进行胶回收，将目的片段和表达载体线性片段 16℃ 连接过夜，体系（10 μL）：Solution I 5 μL；p1301 超量表达载体回收片段 1 μL；目的基因回收片段 4 μL。将连接产物转化大肠杆菌，进行涂板和检测、测序、质粒提取获得融合质粒。酶切体系参照试剂说明书。

表 2-11 表达载体特异引物序列

基因名称	引物序列（5'-3'）
CmERF	F：GGGGATCCCTTTCCTTGGTGAAAAATGGTGAAG R：GGGGTACCATCAGATTGTGGAGCTTCTGCTGA

注：F 为正向引物；R 为反向引物。

（2）叶盘法转化烟草　参考闫朝辉（2017）对本氏烟草的转化。

5.2.6　烟草接种白粉菌后抗性分析

（1）试剂配制　1 mg/mL DAB 染色液，pH 3.8（现用现配）；75% 乙醇；台盼蓝染液：甘油 10 mL、乳酸 10 mL、苯酚 10 mL、台盼蓝（Trypan blue）10 g、无水乙醇 60 mL、无菌水定容至 100 mL；甲基蓝染液：甲基蓝∶甘油∶乳酸∶蒸馏水＝1∶1∶1∶10。

（2）DAB 染色　取处理过的烟草叶片浸入 1 mg/mL DAB 染色液中，尽量减少机械损伤，抽真空 20 min，孵育过夜。取出叶片放入 75% 乙醇溶液中沸水浴脱色 10 min，取出后放在 95%（V/V）乙醇中贮存并在立体显微镜下观察。若为阳性，材料处理部分变成棕色。

（3）台盼蓝染色　取烟草叶片置于台盼蓝染色液中沸水浴 2 min，室温染色 3～4 h，75% 乙醇脱色 10 min。

5.2.7　烟草接种青枯菌和疮痂菌后抗性分析

（1）注射法接种菌液　取保存的青枯菌和疮痂菌液分别在 NA 培养基上划线，28℃ 倒置暗培养 24～48 h。挑取青枯菌和疮痂菌菌落于液体培养基中，28℃，180 r/min，振荡培养。取摇好的菌液 40 mL 于 50 mL 离心管中瞬离，弃上清，用悬浮液悬浮（青枯菌用氯化镁悬浮，疮痂菌用蒸馏水悬浮）至终浓度为 10^8 CUF/mL。用去掉针头的 1 mL 注射器吸取菌液，从叶背面将菌液注射到离体叶片侧叶脉间的叶表面，直至形成水渍状，每片叶接种 4 个部位。用吸饱水的脱脂棉包裹住叶柄，放入带盖塑料盒中保湿，光照/黑暗（16 h/8 h），28℃培养，其间注意及时补充水分。

（2）涂板法测菌液浓度 用直径为 1 cm 的圆形打孔器取接种青枯病和疮痂病部位叶片称重，加无菌水研磨（0.1 g 叶片加水 0.9 mL 为 10 倍稀释液），转入离心管中。在超净工作台中将研磨液稀释（取 10 倍稀释液 0.1 mL 加入 0.9 mL 无菌水中即为 100 倍稀释液，1 000 倍、10 000 倍稀释液依次类推），取 0.1 mL 稀释液涂板，每个稀释度设置 3 个重复，30～300 个菌落为有效板。取 1 000、10 000、100 000 倍的稀释液涂板，28℃倒置培养。计算公式：菌液浓度＝菌落平均数×10×稀释倍数/g。

5.2.8 数据分析

实时定量每个试验数据均为 3 次重复的平均值。采用 Microsoft Excel 2003 和 SAS（SAS Institute，version 8.2）对数据进行分析处理。

5.3 结果与分析

5.3.1 南瓜 CmERF 的克隆及生物信息学分析

（1）南瓜 CmERF 全长的获取 采用同源克隆法从中国南瓜自交系 112-2 中克隆了 CmERF，根据测序结果和同源序列比对可知获得的基因长度为 697 bp，GenBank 登录号：MH105819（图 2-27）。

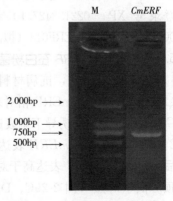

图 2-27 CmERF 基因的克隆

（2）南瓜 CmERF 生物信息学分析 CmERF 基因的 ORF 为

681bp，编码 226 个 aa，分子量为 24.8 ku，等电点为 4.92。氨基酸序列同源比对结果表明（图 2-28），CmERF 编码的氨基酸序列与苦瓜 ERF、甜瓜 ERF、黄瓜 ERF 和拟南芥 ERF 同源率达到 71.98%，并且都含有 ERF 蛋白的 AP2 保守结构域。

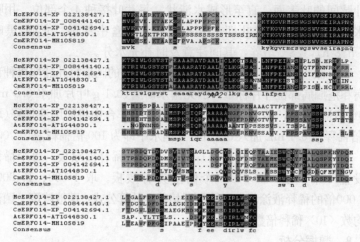

图 2-28 CmERF 蛋白序列比对

红色下划线标出的部分为 ERF 蛋白的 AP2 保守结构域；McERF014（苦瓜，登录号 XP_022138427.1）、CmERF014（甜瓜，登录号 XP_022138427.1）和 CsERF014（黄瓜，登录号 XP_004142694.1）、AtERF014（拟南芥，登录号 AT1G44830.3）。

5.3.2 南瓜 CmERF 在白粉菌和非生物胁迫下的表达分析

白粉菌处理后，抗病材料'九江轿顶'（6 h 除外）和 112-2（12、24 h 除外）mERF 相对表达量低于对照（图 2-29A），表明 CmERF 受白粉病诱导下调表达。SA 处理后，'九江轿顶'和 112-2（3 h 除外）中 CmERF 表达低于对照（图 2-29B）。MeJA 处理后，112-2 CmERF 表达高于对照（0 h、48 h 除外），而'九江轿顶'低于对照（图 2-29C、D）。H_2O_2 处理下，'九江轿顶'和 112-2 中 CmERF 表达低于对照；NaCl、ABA 处理，'九江轿顶'中，CmERF 表达低于对照；112-2 NaCl 处理除 12h、24h 外，均

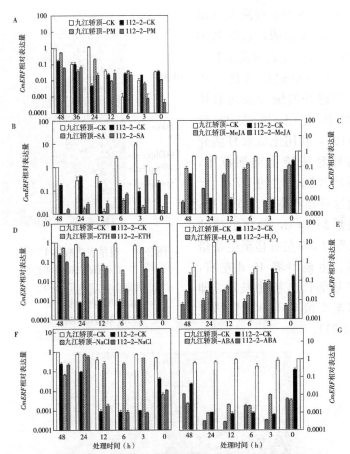

图 2-29 抗/感材料 CmERF 在不同外源处理下的表达分析
A. PM（白粉菌）处理　B. SA 处理　C. MeJA 处理　D. Eth 处理
E. H_2O_2 处理　F. NaCl 处理　G. ABA 处理

低于对照；112-2ABA 处理，除 6h 外，均低于对照（图 2-24E、F、G）。

5.3.3　超量表达烟草的获得

（1）超量表达载体的构建　将扩增到的目的基因片段，连接到 BamHⅠ和 KpnⅠ的酶切过的 P1301 表达载体上，转化大肠杆菌感受态细胞，通过菌液 PCR 筛选出阳性克隆单菌落，提取质粒进行

BamHI 和 KpnI 双酶切（图 2-30），结果表明，目的基因片段均已连接至表达载体上。

（2）烟草遗传转化　将已经构建好的融合表达载体分别转化土壤农杆菌 GV3101，利用土壤农杆菌介导的叶盘法转化至烟草中，烟草叶片脱菌后约 15 d，叶盘边缘开始有愈伤组织出现；脱菌后约 40 d，分化出的幼苗长至 2 cm 左右移至生根培养基。生根培养 40 d 左右，烟草基部长出粗壮根系，即可移至灭菌基质中驯化移栽（图 2-31）。

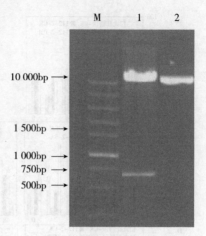

图 2-30　超量表达载体 P1301-CmERF 双酶切检测
泳道 1：P1301-CmERF 双酶切产物；泳道 2：P1301 双酶切产物；M：DL 10 000 DNA Marker

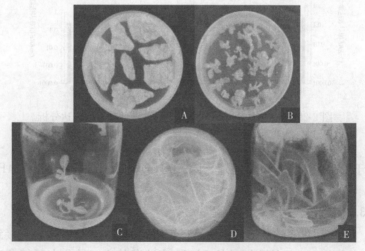

图 2-31　烟草遗传转化
A. 脱菌培养 15 d　B. 脱菌培养 40 d　C. 转移至生根培养基　D、E. 生根培养 40 d

5.3.4　*CmERF* 超量表达对烟草白粉病过敏性反应分析

CmERF 转基因植株叶片接菌处理 1 d 出现的棕色斑点面积和颜色与野生型相差不大，表明 H_2O_2 开始积累；处理 3 d 后，棕色区域颜色面积比对照大但颜色比对照浅；处理 5 d 后，棕色的区域的颜色比对照深，面积比对照大；与接菌 3 d 相比棕色区域的颜色加深，面积增大（图 2-32A）。结果表明转基因植株 *CmERF* 的超量表达可以诱导 H_2O_2 的积累。*CmERF* 转基因植株叶片接种白粉菌后 1 d，蓝色斑点的数量与对照相比明显增加，转基因植株叶片接菌处理 4 d 后出现的蓝色斑点比对照多，而对照处理 7 d 后才出现较多的蓝色斑点，并且处理后 7 d 的转基因植株蓝色斑点与对照相比面积逐渐扩大，颜色加深（图 2-32B）。结果表明在白粉菌处理下，转基因植株 *CmERF* 的超量表达加速了细胞死亡，细胞膜的通透性增加。

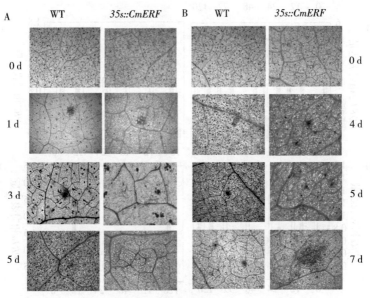

图 2-32　白粉菌处理下烟草叶片 DAB、台盼蓝染色
A. DAB 染色　B. 台盼蓝染色

5.3.5 *CmERF* 超量表达烟草白粉菌胁迫下信号转导相关基因表达分析

以 WT 清水处理 0 h 为对照,分析 *CmERF* 的超量表达烟草及 WT 进行白粉菌接种后信号转导相关基因的表达模式。接种白粉菌后,*PAL* 在转基因烟草中 120 h 的表达量最高,约为清水对照的 5 倍,在 WT 中 0~48 h 表达量逐渐下降,后逐渐升高(图 2-33A),

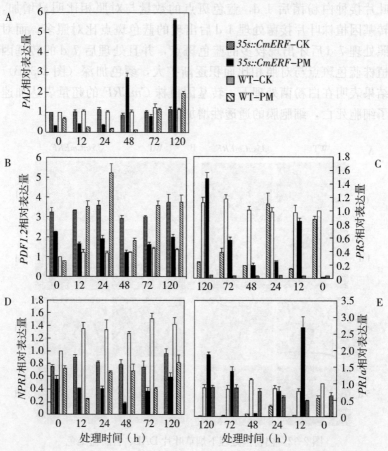

图 2-33 烟草接种白粉菌信号转导基因的表达(烟草 *EF-1α* 作为内参)

结果表明 $CmERF$ 的超量表达正调控 PAL 的表达。接种白粉菌后，$PDF1.2$ 在转基因烟草中稳定表达，WT 中表达增加（图 2-33B），表明 $CmERF$ 的超量表达抑制 $PDF1.2$ 表达。接种白粉菌后，转基因烟草中 $PR5$ 的表达量在 120 h 时达到最高，约为清水对照的 3.3 倍，且在 24～120 h 表达量升高，在 WT 中表达量均下降（24 h 除外）（图 2-33C），结果表明 $CmERF$ 的超量表达正调控 $PR5$ 的表达。接种白粉菌后，$NPR1$ 在转基因烟草和 WT 中表达模式相似（图 2-33D），表明 $CmERF$ 的超量表达对 $NPR1$ 的表达无影响。接种白粉菌后，$PR1a$ 在转基因烟草中 12 h 的表达量最高，约为清水对照的 5.6 倍，24～120 h 表达量逐渐升高，WT 中 $PR1a$ 的表达量最高逐渐降低（图 2-33E），结果表明，$CmERF$ 的超量表达可正调控 $PR1a$ 的表达。

5.3.6 $CmERF$ 超量表达烟草对青枯病和疮痂病的抗性分析

$CmERF$ 转基因烟草第 6 片和第 12 片叶接种烟草青枯菌后，接种部位褪绿变黄程度明显比对照严重。测定其菌液含量发现青枯菌的含量明显高于对照，转基因烟草第 6 片和第 12 片叶的菌液含量分别为对照的 4.65 倍（图 2-34A 和 C）和 5.78（图 2-34B 和 D）倍。这表明 $CmERF$ 超量表达，烟草对烟草青枯病的抗性减弱。转基因烟草第 6 片和第 12 片叶接种烟草疮痂菌后，接种部位褪绿变黄程度明显比对照严重。测定其菌液含量发现疮痂菌的含量明显高于对照，$CmERF$ 转基因烟草第 6 片和第 12 片叶的疮痂菌含量分别为对照的 9.69 倍（图 2-35A 和 C）和 5.75 倍（图 2-35B 和 D）。这表明 $CmERF$ 超量表达，烟草对烟草疮痂病的抗性减弱。

5.4 讨论

ERF 转录因子家族是参与抗病反应的四大类之一，ERF 蛋白在数量上仅次于 MYB 家族。采用同源克隆法从受白粉病胁迫的中国南瓜自交系 112-2 中分离获得南瓜 $CmERF$，编码的氨基酸序列与苦瓜、黄瓜、甜瓜及拟南芥 ERF 基因编码的氨基酸序列具有较

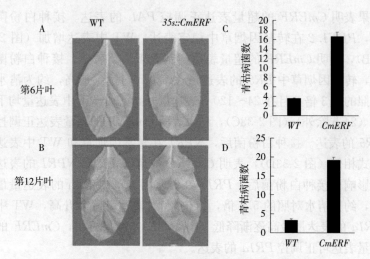

图 2-34 *CmERF* 超量表达对烟草青枯病的抗性分析
A. 烟草第 6 片叶接种青枯病菌后症状
B. 烟草第 12 片叶接种青枯病菌后症状
C、D. 烟草第 6 片叶和第 12 片叶接种部位细菌个数统计

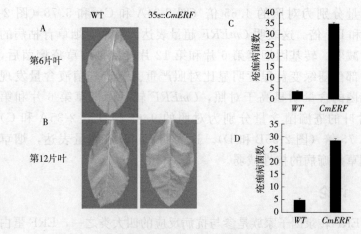

图 2-35 *CmERF* 的超量表达对烟草疮痂病的抗性分析
A. 烟草第 6 片叶接种疮痂病菌后症状
B. 烟草第 12 片叶接种疮痂病菌后症状
C、D. 烟草第 6 片叶和第 12 片叶接种部位细菌个数统计

第2章 南瓜抗白粉病相关基因的分离及功能研究

高的同源率，且具有典型 AP2 保守结构域，表明 CmERF 转录因子基因为 AP2/ERF 基因家族成员。伴随着物种的进化过程，AP2/ERF 转录因子经过了复杂的功能分化过程，在抵抗非生物和生物胁迫等逆境过程中发挥着非常重要的作用。在抗病和感病材料中均存在防御反应基因，但在时间、空间和含量上存在差异，目前普遍认为，防御反应基因被诱导的时间越早，诱导程度越大，其在抗病反应中发挥的作用也越大（Kuc et al., 1985）。

ERF 基因的转录受低温、干旱、病原菌侵染、SA、JA 和 Eth 信号的诱导。从烟草 cDNA 文库中筛选到一个胁迫诱导的 ERF 基因 Tsi1，受高盐、Eth 和 SA 诱导，转基因烟草增强了对烟草野火病（*Pseudomonas syringae* pv. *tabaci*）和高盐的抗性，同时激活了 PR 基因的表达（Park et al., 2001）。外源胁迫处理发现，白粉菌、SA、H_2O_2、NaCl、ABA 处理抑制抗/感材料中 CmERF 表达，MeJA 和 Eth 诱导抗性材料 CmERF 表达，抑制感病材料 CmERF 表达。这些结果表明，CmERF 参与南瓜对白粉菌抗性反应的过程可能是通过 Eth 和 JA 信号转导途径参与南瓜的抗白粉病防御机制。本研究中，CmERF 的超量表达促进 H_2O_2 的积累，细胞膜的通透性增加，加速了细胞坏死，表明 CmERF 基因可能通过调控过敏性反应提高烟草对白粉病抵抗。在烟草中超量表达的 CmERF 基因对烟草青枯病和疮痂病没有抗性，且转基因烟草黄化程度和含菌量均高于 WT，表明 CmERF 的超量表达可能增强烟草对青枯病和疮痂病的感病性。同一种材料在受到不同病原菌侵染时抗病性也会有不同。受到小麦白粉菌（*Blumeria graminis*）诱导，TaERF3 的表达量，抗病材料'Pm97034'高于感病材料'Wan7107'；而受到禾谷镰刀菌（*Fusarium graminearum*）和小麦纹枯病（*Rhizoctonia cerealis*）侵染，TaERF3 基因的表达量，抗病材料'Pm97034'比感病材料'Wan7107'低（Zhang et al., 2004）。ERF 基因在植物中的许多抗病反应是通过信号转导途径来实现的。为了进一步研究 CmERF 的抗病机理，进行信号转导相关基因表达分析。研究表明，在白粉菌胁迫下，CmERF 基因的超

量表达诱导 *PAL*、*PR1a*、*PR5* 上调表达，抑制 *PDF1.2* 表达。说明 *CmERF* 基因可能通过参与 SA 和 JA 信号转导途径来提高烟草对白粉病的抗性。

5.5 小结

从中国南瓜自交系 112-2 中分离白粉病相关基因 *CmERF*，对南瓜进行白粉菌胁迫和非生物胁迫，发现 *CmERF* 可能主要参与 Eth 和 JA 抗病信号转导途径。对转基因烟草进行过敏性反应分析，发现 *CmERF* 的超量表达可增加活性氧的积累，参与烟草对白粉病的抗性；接种烟草青枯菌和疮痂菌，发现 *CmERF* 的超量表达可提高烟草对两种细菌性病原菌的感病性。对抗病信号转导相关基因的表达分析发现，*CmERF* 基因可能是通过调控 *PAL*、*PR1a*、*PR5* 和 *PDF1.2* 基因的表达来提高烟草对白粉病的抗性。

第3章 辣椒对低温胁迫的响应与其低温抗性相关基因的克隆和功能分析

1 文献综述

1.1 茄科植物响应低温胁迫的研究进展

在各种非生物逆境胁迫中，低温严重影响植物的生长发育和地理分布。低温逆境包含冷害和寒害，冷害通常指水分子未结冰，0～15℃对作物造成的伤害；寒害是指水分子结冰，零度以下对植物造成的胁迫（尹延旭，2009）。

辣椒（*Capsicum annuum* L.）属于茄科，是重要的蔬菜和辛辣香料作物，是全球消费量位居第二的蔬菜（Mateos et al.，2008），含有很多人类需要的营养成分，特别是维生素、矿物质和类胡萝卜素等。有些品种含有大量的辣椒素，其药理作用备受关注。在中国，年均种植面积约150万hm^2，已被认为具有高经济价值的作物。2008年1月中国大范围持续低温，雨雪冰冻灾害造成农作物受灾严重，其中蔬菜类受害面积最大，受害程度严重。据报道，江门市种植的辣椒由于寒冷造成花朵死亡，无法挂果，受灾面积7 310亩，共损失约420万元。因此，提高辣椒的抗逆性，挖掘辣椒抗逆基因，剖析辣椒抗逆基因的作用机理，培育抗逆辣椒品种具有十分重要的现实意义。

低温主要引起膜损伤，部分归因于植株的急剧失水。低温信号被细胞膜感受器感知，然后转导开启低温相关的基因和转录因子。为了克服低温造成的伤害，植物诱发一系列信号转导，引起基因表达改变,从而调整生理生化代谢来适应逆境(Smallwood,Bowles,2002;Zhu et al.,2007)。参与低温适应性的基本机制主要包括细胞膜脂改变,细胞内钙离子的积累、活性氧(ROS)的增加和 ROS 清除系统的启动、低温抗性相关基因和转录因子表达的改变及蛋白、糖类合成和脯氨酸积累等。

1.1.1 生理生化方向的研究进展

(1)细胞膜系统与抗寒性 早在 20 世纪 70 年代,Lyons 和 Raison(1970)认为细胞膜系统是植物遭受低温伤害的首要部位,提出了著名的"膜脂相变学说",其主要内容:低温袭击植物时,首先引起膜相改变,由液晶相变为凝胶相,引发无序脂肪链变为有序,导致膜破裂和膜结合相关的酶结构发生改变。

膜相转换的温度与膜脂肪酸的成分有关,即饱和脂肪酸成分越高,膜相转换的温度也就越高,不饱和脂肪酸含量越高,膜脂的相变温度越低。Wada et al.(1994)利用抗冷蓝绿藻克隆了与膜脂肪酸不饱和度有关的基因 *desA*,并将其导入不耐冷蓝绿藻,结果膜脂组成发生改变,在低温 5℃时光合作用受抑制程度减轻。膜脂脂肪酸的去饱和是调控植物耐寒性的一个重要机制,为提高植物的耐寒性提供了一个有效的研究方向。

(2)活性氧清除系统与低温 低温引起 ROS 积累,导致植物的膜脂过氧化(苏维埃,1998)。膜脂过氧化是自由基对膜脂不饱和脂肪酸双键引发的一系列反应,最终产生对细胞有害的脂质过氧化物(陈娜等,2005),其主要产物为丙二醛。植物体内代谢产生的 ROS 种类主要包括双氧自由基($^3O_2$)、超氧阴离子(O_2^-)、氧阴离子(O^{2-})、单线态氧($^1O_2$)、过氧化氢自由基($\cdot HO_2$)、羟基自由基($\cdot HO$)、过氧化氢(H_2O_2)等(Apel and Hirt 2004)。在非胁迫条件下,植物体内活性氧的代谢处于平衡状态,对植物不能造成伤害(Elstner,1982);但当植物受到非生物逆境胁迫时,植物体内产生大量的活性氧,如果体内防御系统不足以清除过量的活性氧就会对植物造成伤害(刘鸿先等,1985)。为了防止过量活性氧的积累,植物形成对其清除的生理生化反应机制,包括抗氧化酶和抗氧化物。抗氧化酶主要由以下组成：超氧化物歧化酶（SOD）、过氧化氢酶（CAT）、过氧化物酶（POD）和抗坏血酸过氧化物酶（APX）、谷胱甘肽还原酶（GR）、脱氢抗坏血酸还原酶（DHAR）、单脱氢抗坏血酸还原酶（MDHAR）等；抗氧化物是指还原型抗坏血酸（AsA）、谷胱甘肽（GSH）、脯氨酸（Pro）和β-胡萝卜素（β-carotene）等（Fryer et al., 1998；Prasad，1997）。

遭受低温胁迫黄瓜幼苗SOD、POD和CAT活性均明显增强，并保持稳定水平（李明玉等，2006）；大量报道辣椒叶片POD和SOD活性在遭受低温胁迫时缓慢增加（马艳青等，2000；任旭琴等，2006；杨广东和郭庆萍，1998；邹志荣和陆帼一，1994）。遭受冷胁迫时，耐冷番茄伤害和膜脂过氧化程度小于冷敏番茄，耐冷自交系SOD、CAT和POD酶活性的变化率小于冷敏自交系（郑东虎等，2003）。刘鹏等（2003）研究发现，与未锻炼苗相比，冷锻炼有效地提高了甜椒幼苗叶片的膜保护系统，遭受低温胁迫时表现出较高的SOD酶活性。在低温和干旱胁迫环境下，植物体内APX、MDHAR、DHAR及GR活性在清除H_2O_2等活性氧系统中发挥重要作用（Jin et al., 2003）。高俊杰等（2009）研究表明遭受低温胁迫时，嫁接苗DHAR和GR酶活性及相关基因表达高于自嫁苗，AsA和GSH含量与AsA/DHA和GSH/GSSG比值也较高。

1.1.2 分子方向的研究进展

（1）低温信号转导基因

①钙信号转导与低温。植物细胞通过膜僵化效应感知低温胁迫。低温条件下质膜的僵化诱导低温响应基因，有助于冷适应（Orvar et al., 2000；Sangwan et al., 2001）。拟南芥突变体 *fad2*（缺少油酸去饱和酶）表现膜僵化和低温敏感，表明低温胁迫的感知与质膜僵化有关（Vaultier et al., 2006）。另外，植物遭受低温胁迫时胞质钙离子增加。钙离子在调节植物生长、发育和响应环境刺激时传递信号（Du and Poovaiah, 2005；Sanders et al., 2002）。低温诱导的质膜僵化引起肌醇骨架重排，伴随着激活钙离子通路和细胞质内钙离

子增加,从而产生冷适应反应(Catala et al.,2003;Sangwan et al.,2001)。在低温信号转导过程中,钙离子调节 CBFs 转录因子和 *CORs*(低温响应基因)表达(Chinnusamy et al.,2007,2010)。Doherty et al.(2009)证明了钙信号与 CBFs 转录因子的低温诱导有关,钙调素结合转录启动子(CAMTA)与 *CBF2* 基因启动子的调控元件结合。CAMTA 蛋白是钙调素结合转录因子,直接把低温诱导的细胞质内钙信号传递到下游基因。

②ROS 与低温。ROS 在非生物胁迫中所起的作用是个值得深入研究的问题,特别是 ROS 被报道参与植物的逆境适应过程(Suzuki et al.,2011)。ROS 不仅是植物代谢的有毒副产物,而且作为信号分子介导各种基因表达,如编码抗氧化酶和 H_2O_2 产生的调节子(Neill et al.,2002;Gechev et al.,2003;Suzuki et al.,2011)。低温诱导下 AsA 和 GSH 含量增加,主要引起 NADpH 生成相关的脱氢酶活性提高(Airaki et al.,2011);另外在低温条件下,ROS 激活一些激酶进行信号放大传递,引起植物的低温适应性(Kovtun et al.,2000;Teige et al.,2004)。

(2) 低温响应基因和转录因子 在逆境胁迫下,各种基因的表达发生改变。这种逆境相关基因从两个方面保护植物。一方面产生重要的代谢蛋白,另一方面逆境相关基因的产物参与信号转导途径基因的调节。根据这些逆境相关基因编码蛋白的功能,将其分为两类(Fowler and Thomashow,2002;Kreps et al.,2002;Seki et al.,2002)。一类是在抗逆中可能起作用的蛋白:晚期胚胎丰富蛋白、渗透蛋白、抗冻蛋白、脯氨酸和抗氧化酶等;另一类是参与信号转导和基因调节的蛋白,如各种转录因子(Seki et al.,2003)。

①CBFs 转录因子。在低温条件下,低温响应基因被诱导表达。甜椒 2 个 CBFs 转录因子(*CaCBFIA* 和 *CaCBFIB*)受低温诱导(Kim et al.,2004)。植物遭受低温胁迫时也能诱导 CBFs 转录因子基因如拟南芥 *AtCBF1*(*DREB1B*)、*AtCBF2*(*DREB1C*)和 *AtCBF3*(*DREB1A*)表达,表明在低温条件下,这些转录因子调节低温响应基因的表达。这些转录因子的超量表达不仅能够提

高植株的耐寒性，而且抗旱性也得到提高（Liu et al.，1998）。转基因拟南芥组成型或逆境诱导的 CBF1/3 转录因子超量表达增强低温响应基因的组成表达，从而提高抗逆性（Liu et al.，1998；Jaglo-Ottosen et al.，1998；Kasuga et al.，1999）。且拟南芥 CBF1/3 转录因子的超量表达也能够提高其他植物的抗逆性，如番茄（Hsieh et al.，2002）、烟草（Kasuga et al.，2004）。然而，*CBFs* 转基因拟南芥的转录组表达分析表明仅有 12% 低温响应基因受 CBFs 转录因子诱导表达（Fowler and Thomashow，2002）。这表明在低温适应过程中有其他的转录启动子或抑制子存在。

②与抗氧化防御相关的酶基因。过量表达 *Cu/ZnSOD* 和 *APX* 能够提高除草剂甲基紫精（MV）诱导的抗氧化胁迫能力，如烟草（Kwon et al.，2002）和马铃薯（Tang et al.，2006）。CAT 反义 RNA 转基因番茄植株 CAT 活性下降，增加了对氧化胁迫的敏感性，表明 CAT 在番茄的抗氧化清除系统中起着重要的作用（Kerdnai-mongkol，Woodson，1999）。将拟南芥 C-脱水应答素结合因子 1（*CBF1*）转入番茄，CAT 活性显著增加，植株的耐冷性得到提高（Hsieh et al.，2002）。在棉花叶片上过量表达抗氧化酶基因（*SOD*、*APX* 和 *GR*），转基因植株提高光化学光能的利用效率，保持较高的电子传递速率，从而减轻低温诱导产生的 PSII 光抑制（Kornyeyev et al.，2001）。APX 和谷氨酰胺合成酶基因与植物的耐寒性也有关系。热激锻炼诱导水稻 *APXa* 基因表达，使其活性提高，从而提高幼苗的耐寒性（Sato et al.，2001）。在水稻上过量表达叶绿体谷氨酰胺合成酶基因，转基因植株耐寒性得到提高（Hoshida et al.，2000）。然而也有相反的报道，比如：Foyer et al.（1994）研究表明在植物中抗氧化酶 *SOD* 的超表达可能产生正、负效应。Slooten et al.（1995）也报道在叶绿体中 *MnSOD* 的超量表达对低温引发光抑制的抗性没有增强。

1.1.3 ABA 响应低温胁迫反应的研究进展

脱落酸（ABA）是一类重要的逆境激素，介导非生物逆境信号和抗逆性。植物响应非生物逆境胁迫时 ABA 积累，如干旱和盐胁迫（Xiong and Zhu，2003）。低温也能增加内源 ABA 含量，但幅度较小（Lang et al.，1994）。因为 ABA 合成主要在转录水平受调节（Xiong and Zhu，2003）。拟南芥 ABA 合成相关基因不受低温调控，表明在低温胁迫下，植物体内 ABA 合成不占主导（Lee et al.，2005）。然而，ABA 信号转导参与低温反应。低温胁迫诱导很多基因同样也受 ABA 上调表达（Thomashow，1999）。

(1) ABA 介导低温信号反应　ABA 作为信号分子，通过第二信使如磷脂酶 D、过氧化氢和钙离子，在低温信号转导过程中起着重要的作用。磷脂酶 D 与 ROS 有关，ROS 参与 ABA 和低温反应（Laloi et al.，2004；Mahajan and Tuteja，2005）。在拟南芥上，不同类型的磷脂酶 D 如 AtPLDa1 和 AtPLDd 参与 ROS 的产生和响应（Zhang et al.，2003，2005）。*AtPLDd* 受 ABA 和低温诱导表达（Katagiri et al.，2001）。低温诱导的耐冻性在 *AtPLDd* 突变体受阻，而在过量表达株系上得到提高，分别出现冻害诱导的磷脂酸产物下降和增加（Li et al.，2004）。钙离子通过影响磷酸化酶参与植物对 ABA 和低温反应。受体蛋白激酶（RPK1）受 ABA 诱导，并在 ABA 信号转导过程中起着重要的作用（Hong et al.，1997；Osakabe et al.，2005）。RPK1 在响应后期低温反应时上调表达，且参与识别早期低温产生的第二信号（Lee et al.，2005）。这些结果表明低温信号转导与 ABA 信号转导存在共同途径。

(2) ABA 与低温反应下游基因　ABA 介导的低温反应下游基因主要通过两个途径表达。一个是下游基因的表达不依赖 ABA。不依靠 ABA 的基因表达由 bZIP 转录因子调节，属于 MYC 和 MYB 家族。水稻 *OsMYB4*（MYB 家族成员）受低温诱导表达，不受 ABA 上调（Park et al.，2010）。水稻 *MYBS3* 在转录水平抑制依靠 DREB1/CBF 的低温信号转导，而且在低温胁迫过程中，*DREB1* 响应快速而短暂，而 *MYBS3* 反应缓慢（Su et al.，2010）。在拟南芥上过量表达 *CBF1*、*CBF2* 和 *CBF3* 基因，不影响 *MYBC1* 的表达，表明 *MYBC1* 不受这些 CBF 转录因子下游调节（Zhai et al.，2010）。另一个是低温反应下游基因的表达依赖

ABA。ABA 提高抗氧化防御系统，减少低温诱导的活性氧积累（Liu et al.，2011）；而且 ABA 也能诱导 *CBF1*、*CBF2* 和 *CBF3* 转录因子的表达，进而通过 CRT 启动子元件调节低温响应基因（Knight et al.，2004）。拟南芥低温诱导基因 *RAB18* 同样受 ABA 上调表达，但该基因在 ABA 突变型和缺失型拟南芥中不能表达，说明低温诱导的表达依赖 ABA（Lang et al.，1992）。

1.2 逆境相关基因的研究进展

培育转基因耐寒品种是快速和有效改良作物的生物技术方法。转基因技术已经被成功用来提高水稻的耐寒性。分离耐寒相关基因是利用遗传转化改良作物的重要第一步（Dubouzet et al.，2003；Hsieh et al.，2002；Ito et al.，2006；Ohnishi et al.，2005）。在分离的耐寒相关基因中，转录因子可以更好地提高植物的抗逆性。

1.2.1 NAC 转录因子

（1）NAC 转录因子的结构特征　植物特有的 NAC 转录因子是最大的转录因子家族之一。目前发现在模式植物中存在大量的 *NAC* 基因，如拟南芥 117 个、水稻 151 个、葡萄 79 个、杨树 163 个、大豆 152 个、烟草 152 个。典型的 NAC 蛋白含有高度保守的 N 端 NAC 结构域（约 150 个氨基酸）和 C 端多变的转录调节区域。NAC 结构域分为 5 个亚结构域（A、B、C、D、E）。C 和 D 高度保守，与 DNA 结合，A 亚结构域参与功能二聚体形成，B 和 E 序列多变，负责 NAC 基因的功能多样性。NAC 亚结构域（D 亚结构域）富含 Lys 氨基酸的核定位信号。转录调节区域位于高度多样化的 C 端，激活或抑制转录。该区域富含几个简单重复序列如丝氨酸-苏氨酸、脯氨酸-谷氨酸或酸性氨基酸残基。有些 *NAC* 基因 C 端含有 α-螺旋跨膜结构（命名为 NTLs），负责锚定质膜或内质网膜。目前为止，已在拟南芥上发现 18 个 *NTLs* 基因、大豆 11 个、葡萄 6 个、水稻 5 个，它们在响应环境刺激时起着重要的调节作用。

（2）NAC 转录因子在植物生长发育中的作用　植物 NAC 转

录因子在植物的生长发育过程中起着重要的作用,主要参与侧根生长、叶片衰老和次生壁形成等方面。拟南芥 *AtNAC1* 和 *AtNAC2* 的过量表达促进侧根形成(He et al.,2005;Xie et al.,2000)。拟南芥约 20 个 *NAC* 基因的表达与叶片衰老有关。*ANAC092* 过量表达株系有 170 个基因上调表达,其中 78 个与衰老有关(Balazadeh et al.,2010)。*AtNAC* 促进叶片衰老,具有组织表达特异性,仅在衰老的叶片中表达,幼嫩的绿叶中无表达;*AtNAC* 过量表达转基因植株表现叶片过早衰老,而且该基因的缺失突变体叶片表现延迟衰老,水稻和豌豆的同源基因能使延迟表型恢复(Guo et al.,2006)。水稻 *OsNAC5* 基因调节叶片的衰老,在自然衰老和逆境诱导(ABA 和高盐)的衰老过程中,该基因上调表达(Sperotto et al.,2009)。

有些 *NAC* 基因对细胞次生壁的形成起着促进作用,如 *SND1* (Secondly wall-associated NAC domain 1) 和 *VND6/7* (Vascular-related NAC domain 6/7)。过量表达 *SND1* 株系激活次级细胞壁生物合成相关基因的表达,使细胞壁变厚;反义表达该基因的株系表现纤维次级壁厚度显著下降(Zhong et al.,2006)。*VND6* 和 *VND7* 诱导细胞转分化为原生木质部和后生木质部导管。过量表达 *VND6* 和 *VND7* 株系表现原生木质部和后生木质部导管的加速形成,而抑制其表达时,株条根部原生木质部和后生木质部导管的形成同样也受到抑制(Kubo et al.,2005;Yamaguchi et al.,2008)。此外,也有一些 *NAC* 基因在细胞次生壁的形成中起着负调节作用,如 *ANAC012* 和 *XND1* (xylem NAC domain 1)。*XND1* 过量表达株系下胚轴原生木质部组织的薄壁细胞缺乏次生壁增厚能力,植株出现矮化和木质部导管缺失现象;敲除 *XND1* 株系,导管长度变短,植株也矮小(Zhao et al.,2008)。

(3) NAC 转录因子在响应非生物逆境胁迫中的作用 近几年研究表明很多 *NAC* 基因参与非生物逆境胁迫反应,并在响应过程中起着重要的调节作用。拟南芥 *AtAF1* 受干旱和 ABA 诱导表达,T-DNA 插入突变体 ataf1-1 和 ataf1-2 恢复率是野生型的 8 倍,具有较高的抗旱性,并且上调一些逆境相关基因表达如 *COR47*、*ERD10*、*KIN1*、*RD22* 和 *RD29A*(Lu et al.,2007)。拟南芥 *RD26* 受干旱、高盐和 ABA 诱导表达,*RD26* 过量表达株系对 ABA 高度敏感,而且诱导表达一些 ABA 和非生物逆境相关的基因,从而增强植株的抗旱性;而敲除 *RD26* 则表现对 ABA 不敏感,并且相应的 ABA 效应基因也受到抑制(Fujita et al.,2004)。水稻 *OsNAC6* 受低温、干旱、高盐、ABA 和 SA 诱导表达,过量表达 *OsNAC6* 株系遭受脱水、高盐和稻瘟病菌胁迫时表现较强的抗性(Nakashima et al.,2007),或许与生物和非生物胁迫相关基因上调表达有关。干旱、高盐、低温和 ABA 也可诱导水稻 *OsNAC045* 表达,过量表达 *OsNAC045* 转基因株系增强对干旱和高盐胁迫的防御能力,而且诱导胁迫相关基因 *OsLEA3-1* 和 *OsPM1* 表达。拟南芥 *ANAC102* 受 0.1% 低氧处理诱导表达,且该基因表达下调使种子发芽率显著降低,而增加其表达对种子萌发没有影响;过量表达株系诱导很多与低氧敏感相关的基因,植株叶片略微变黄(Christianson et al.,2009)。另一拟南芥基因 *NTL8* 通过 GA 信号途径参与盐胁迫种子萌发,受高盐诱导表达,其表达不依赖 ABA。*ntl8-1* 突变体种子萌发对高盐胁迫的抗性增强(Kim et al.,2008)。

1.2.2 植物 MBF1 转录辅激活因子

(1) 转录辅激活因子的作用机制 核受体是一类细胞核内或与相应配体结合后转移到核内的受体。核受体转录因子活性的激活需要辅助调节蛋白,即辅调节子的参与(McKenna,O'Malley,2002)。辅调节子桥连核受体和基础转录因子而发挥作用(Hermanson et al.,2002)。MBF1 就是一类核受体辅激活因子,通过桥连通用转录因子 TBP 和基因特异性转录因子,促进特异性转录因子的 DNA 结合能力,或通过聚集 TBP 到启动子,从而增强靶基因的转录活性。TBP 识别 TATA 盒并与其结合,是一类存在于真核细胞中的通用转录因子。

(2) 植物 MBF1 在生长发育和逆境胁迫下的研究进展 植物

MBF1 参与生长发育和环境胁迫反应，研究较多的是模式植物拟南芥。拟南芥中有 3 个 *MBF1* 基因，即 *AtMBF1a*、*AtMBF1b* 和 *AtMBF1c*，这 3 个基因在不同组织中表达模式不同（Tsuda and Yamazaki，2004）：*AtMBF1a* 仅在花药和种子中表达，*AtMBF1b* 在叶脉、茎、花药和种子中均有表达；*AtMBF1c* 在所有组织中都表达，包括花和果荚。

拟南芥 MBF1 的功能与植物的生长变化有关，如植株大小、叶片细胞宽度（Hommel et al.，2008；Tojo et al.，2009）。*AtMBF1s* 通过负调节 ABA 抑制种子萌发（Mauro et al.，2012）。抑制 *AtMBF1c* 表达转基因植株茎伸长明显迟缓，植株矮小，叶柄和叶片长度显著减短，而且转基因植株的种子萌发率大幅降低（Hommel et al.，2008）。*AtMBF1c* 超量表达转基因拟南芥表型比对照野生型大 20%，结籽量增加（Suzuki et al.，2005）。而 *AtMBF1a* 超量表达拟南芥与野生型相比，形态和生长没有发生变化（Kim et al.，2007）。*StMBF1* 结合植物不同发育阶段的 Hd-Zip 蛋白 Hahb-4，进而增加 Hahb-4 和 HAHR-1 的 DNA 结合能力（Zanetti et al.，2006）。

高温和干旱诱导番茄 *LeMBF1*，乙烯对其表达没有影响（王达菲，2009）。烟草 *MBF1* 受高温和干旱同时处理诱导表达（Rizhsky et al.，2002）；ABA 对拟南芥 *AtMBF1a* 和 *AtMBF1b* 表达没有影响，而显著诱导 *AtMBF1c* 表达（Tsuda et al.，2004）；马铃薯 *StMBF1* 受机械损伤、病菌侵染和 SA 强烈诱导表达（Godoy et al.，2001）。盐胁迫诱导拟南芥 *AtMBF1a* 和 *AtMBF1b* 表达（Kim et al.，2006），干旱和高温对 *AtMBF1a* 和 *AtMBF1b* 表达没有影响（Suzuki et al.，2005）。*AtMBF1c* 响应逆境胁迫如高温、过氧化氢、失水和高盐时上调表达（Tsuda and Yamazaki，2004）。拟南芥 *AtMBF1a* 组成型超量表达有助于提高植株的耐盐性、对葡萄糖的不敏感性和抗病性，并进一步证实在响应病原菌侵染时，参与乙烯/茉莉酸信号转导（Kim et al.，2007）。*AtMBF1c* 超量表达转基因植株通过部分激活或扰乱乙烯信号转导提高植物对高温、渗透胁迫和病原菌的抗性；另外 MBF1 超量表达引起很多编码逆境反应相关的转录因子和信号转导基因的积累，如 DREB2A、WRKY 转录因子、钙结合蛋白、锌指蛋白 Zat7 和 Zat12、APX2 和 PRs 蛋白等（Suzuki et al.，2005，2011）。

1.2.3 F-box 蛋白家族

（1）F-box 蛋白的结构特征　泛素化蛋白酶水解系统介导一些蛋白降解，从而在植物的生长发育中起着重要的作用。泛素化途径包含 E1 泛素激活酶、E2 泛素转化酶和 E3 连接酶。E3 连接酶最终把泛素多聚体连接到底物上，进而介导底物降解。E3 连接酶负责识别底物的特异性，分为几类：HECT、APC、VBC-Cul2、Ring/U-box 和 SCF。SCF 复合物是 E3 连接酶最重要的一类，由 Skp1、Cul1/Cdc53、Roc1/Rbx1/Hrt1 和 F-box 蛋白组成。F-box 蛋白在 N 端含有高度保守的 F-box 结构域，C 端负责与多样的底物结合。F-box 结构域在介导蛋白与蛋白互作中起着重要的作用，如泛素多聚化、转录延伸和抑制翻译。在泛素蛋白酶水解系统中，F-box 结构域连接 F-box 蛋白和 SCF 复合物的其他成员如 Skp1 或 Skp1-like 蛋白。F-box 蛋白的 C 端特异性结合底物，含有最具典型的 motifs 如 LRR（亮氨酸的简单重复序列）、Kelch 重复序列和 WD-40 重复序列，但是大部分 F-box 蛋白含有未知的 motifs。

（2）F-box 蛋白在植物生长发育中的作用　拟南芥 F-box 蛋白（UFO）是花愈伤组织形成所必需的，在花芽分化过程中起着重要的作用，参与花器官的轮生形成，控制花愈伤组织的分化，以及激活花瓣和雄蕊形成所必需的 *APETALA3* 和 *PISTILLATA* 基因表达；*UFO* 突变体抑制花器官形成，在花器官生长过程中控制细胞增殖，产生纤维组织（Levin，Meyerowitz，1995）。水稻 F-box 蛋白基因 *MAIF1* 被 ABA 和非生物逆境快速和显著诱导表达，根尖部位 *MAIF1* 也被蔗糖、生长素和细胞分裂素诱导表达；过量表达株系降低对 ABA 的敏感性和非生物胁迫的抗性，促进根系生长，表明 *MAIF1* 基因通过调节根系生长而在响应非生物胁迫反应中起着负调控作用（Yan et al. 2011）。拟南芥 F-box 蛋白（ORE9）突变体在自然衰老中延缓各种衰老症状，而且在激素诱导的衰老中也表现为延迟衰老。这些结果表明 ORE9 能够缩短叶龄，通过延缓叶片衰老所需的蛋白降解发挥作用（Woo et al., 2001）。

(3) F-box 蛋白与植物激素信号转导　F-box 蛋白在很多信号转导途径中发挥重要作用。在乙烯信号转导中，拟南芥 2 个 F-box 蛋白（EBF1 和 EBF2）使转录激活子 EIN3 降解，进而负调节乙烯反应（Guo, Ecker, 2003）。拟南芥 EIN3 蛋白是介导乙烯调节的基因表达和形态反应的关键转录因子，组成型表达，是仅与乙烯识别就可以结合目标基因的启动子（Chao et al., 1997）。在缺少乙烯的情况下，EIN3 被 SCF EBF1/2 复合物泛素化降解。相反乙烯存在的情况下，EIN2 阻止 EIN3 泛素化，引起 EIN3 积累，从而启动参与乙烯反应的基因表达（Guo and Ecker, 2003; Potuschak et al., 2003）。EIN3 作为转录激活子在乙烯信号转导中被降解，不同于 Aux/IAAs 作为生长素反应的抑制基因被降解。拟南芥 F-box 蛋白 TIR1 是生长素受体（Dharmasiri et al., 2005; Kepinski, Leyser, 2005），Aux/IAAs 和 ARFs 是生长素介导基因的关键调控子，ARFs 是一类与启动子特异结合的转录因子（Ulmasov et al., 1999）。Aux/IAAs 和 ARFs 结合通过抑制区域负调控 ARF 蛋白的转录激活区域（Tiwari et al., 2004）。生长素通过 SCF TIR1 促进 Aux/IAAs 泛素化进行降解，从而释放出 ARFs，ARF 形成聚合体进行介导生长素诱导的基因表达（Moon et al., 2004; Smalle et al., 2004）。在生长素和 GA 信号转导中，F-box 蛋白都是通过 SCF 复合物降解负调节因子而参与其信号转导。作为重要的调节因子，DELLA 负调控 GA 信号转导（Fleet et al., 2005; Thomas et al., 2004），F-box 蛋白 SLEEPY1（SLY1）使 DELLA 降解进而介导 GA 信号反应（Strader et al., 2004）。F-box 蛋白（COI1）是 JA 信号转导途径中的重要调节因子，所有依赖 JA 的反应都需要 COI1 的参与（Bai et al., 1996; Feys et al., 1994）。

1.2.4 MADS-box 转录因子

MADS-box 基因编码一类重要的转录因子,参与调节真核生物的生长发育和信号转导。MADS-box 结构域高度保守,含有约 58 个氨基酸,具有与 DNA 结合、多聚化和核定位特征。在拟南芥、水稻和番茄中分别鉴定出 132、97 和 36 个 MADS-box 基因。植物花器官发育过程中的大多数基因都属于 MADS-box 家族基因,该家族基因的功能在调控开花时间、愈伤组织分化和花器官形成方面的研究较为深入。另外,MADS-box 基因还参与果实成熟、胚胎发育和营养器官形成(如根和叶)。绝大多数与植物的生长和发育有关,仅有小部分参与非生物逆境胁迫。拟南芥 AGL91 受低温诱导表达。水稻 4 个 MADS-box 基因(OsMADS18、22、26 和 27)在响应低温和失水胁迫反应时上调表达;另外 3 个 MADS-box 基因(OsMADS2、30 和 55)响应失水和高盐胁迫时下调表达。番茄 MADS-box 基因的表达受低温显著诱导,且表达的变化与花发育畸形有关。玉米 MADS-box 基因(ZMM7-L)在响应低温、高盐和 PEG 处理过程中上调表达,而外源 ABA 处理抑制其表达;在失水胁迫下过量表达株系发芽率低于野生型,表明 ZMM7-L 基因可能是非生物胁迫反应中的负调控因子。

1.2.5 脱水素(DHN)

脱水素属于后期胚胎丰富蛋白 D-11 亚家族,在耐干燥种子胚胎或者遭受各种环境胁迫的营养组织中大量积累,如干旱、高盐和低温。DHN 的结构特征是具有保守性的 K、S 和 Y 结构。K 区域存在于所有的 DHN 基因,由高度保守的 15 个氨基酸组成(EKKGIMDKIKEKLPG),进而形成具有两性分子的 α-螺旋。根据这些结构域的排布,将 DHN 基因分为 4 类,分别是 YnSKn、Kn、KnS 和 SKn。

拟南芥脱水素基因(COR47、RAB18、LTI29 和 LTI30)受低温诱导。并且在植物不同组织中不同种类的 DHN 基因被不同的环境因子诱导表达,COR47 和 LTI30 主要在微管组织中表达,受低温诱导,而 ERD14 和 LTI29 在正常生长条件下的根尖中表达。番茄 DHN 基因（TAS4）受渗透胁迫和 ABA 诱导表达。在正常生长条件下，过量表达 TAS4 株系增强对长期干旱和高盐胁迫的抗性，且没有影响植株生长。DHN 基因受非生物胁迫显著诱导，如干旱、低温和高盐，表明 DHN 基因表达与抗逆性存在正相关关系，并且过量表达株系的抗逆性增强。BjDHN3 转基因烟草经 4℃低温处理后，耐寒性明显提高。

1.3 研究的主要内容

（1）外源 ABA 对低温下辣椒幼苗抗氧化酶活性及其基因表达的影响，主要探讨 ABA 是否启动 AsA-GSH 循环代谢进行活性氧的清除，进而提高辣椒的耐寒性。

（2）从分子水平挖掘 ABA 调控的耐寒相关基因，利用抑制消减杂交技术（SSH 文库）进行差异基因筛选，主要对遭受低温胁迫的辣椒幼苗进行分离 ABA 调节的差异基因。

（3）辣椒低温抗性相关基因克隆及其功能分析。根据 SSH 技术分离的辣椒低温相关 ESTs，从中筛选感兴趣的差异片段进行克隆全长，采用实时定量 PCR 技术分析非生物胁迫下的表达模式，并且利用 VIGS 敲除技术和转基因过量表达方法对其进行功能分析。

2 外源 ABA 对低温下辣椒抗氧化酶活性及其基因表达的影响

AsA 是植物细胞中重要的抗氧化物，直接清除 ROS 或间接在 APX 还原过氧化氢的过程中提供电子。抗坏血酸-谷胱甘肽循环（AsA-GSH cycle）包括 APX、MDHAR、DHAR 和 GR 酶。以前对循环代谢的研究主要集中在逆境条件下 AsA-GSH 循环涉及的酶活性变化（Selote and Khanna-Chopra, 2006; Stevens, 2008）。MDHAR 活性的增加有利于提高番茄果实的耐寒性（Stevens, 2008）。DHAR 基因的过量表达通过增加 AsA 的含量提高番茄的耐寒性（Le Martret et al.，2011）。然而，到目前为止关于 AsA-GSH 循环的相关研究在园艺作物上报道很少（Li et al.，2010）。

ABA 作为逆境信号分子，能够提高植物的抗逆性，如低温（Verslues et al.，2005）、高盐（Bellaire et al.，2000）和干旱（Ma et al.，2008）。这种抗逆性的提高一方面归因于抗氧化防御系统的增强如 SOD、CAT、APX 和 GR（Bellaire et al.，2000）和 AsA、GSH（Jiang and Zhang，2002），预防活性氧的积累。外源 ABA 通过增加 SOD 活性、AsA 和 GSH 含量来提高柱花草的耐寒性，然而关于 AsA-GSH 循环如何代谢进行清除过量活性氧的报道很少（Zhou et al.，2005）。在拟南芥上，ABA 介导 AsA-GSH 循环相关基因的上调参与活性氧的清除，并且 ABA 在转录水平通过增强这个循环代谢介导 AsA 积累（Ghassemian et al.，2008）。然而，外源 ABA 通过 AsA-GSH 循环能否提高植物耐寒性的研究很少。在辣椒上，关于 ABA 对 AsA 和 GSH 含量的影响及其循环相关的活性氧清除酶如何调节 AsA 再生的报道很少。为此，我们研究外源 ABA 对这些生理指标的影响，如丙二醛（MDA）、过氧化氢（H_2O_2）、常规抗氧化酶活性和 AsA-GSH 循环关键酶活性，并对 *Mn-SOD*、*POD*、*DHAR1* 和 *DHAR2* 基因的表达模式进行实时定量分析。

2.1 材料与处理

供试辣椒耐寒材料 P70 由西北农林科技大学辣椒课题组提供。种子温汤浸种进行催芽，即种子浸在温水（50~60℃）里约 20 min，中间不断搅动。然后放到常温水中吸水 4~6 h，最后取出放到光照培养箱中进行催芽，间隔 1 d 冲洗 2 次，条件设为 25~28℃，相对湿度 60%~80%，黑暗。当发芽率达到 80%，播种到穴盘中，基质用草炭和珍珠岩按 1∶3 混匀。当植株生长 6~8 片真叶时，于早上 9：00—10：00 时采用 0.57 mmol/L ABA 溶液进行叶面喷施，ABA 溶液里添加 0.5% Tween-20 增加表面附着。叶面喷施 3 d 后，进行低温处理，10/6℃（白天/黑夜），光强 5 500 lx。试验设计 4 个处理：对照（CK），喷施清水后进行常温处理；ABA，喷施 ABA 后进行常温处理；低温，喷施清水后进行低温处理；ABA＋低温，喷施 ABA 后进行低温处理。

2.2 试验方法

2.2.1 坏死斑的统计

按照以下级别对植株的冷害程度进行划分。1 级：无症状；2 级：坏死斑面积小于 5%，且不抑制生长；3 级：出现明显的坏死斑（5%～25%）；4 级：大面积的坏死斑且抑制植物生长（26%～50%）；5 级：整株出现坏死斑和倒伏。冷害指数按照公式进行计算：冷害指数＝（级别×某一级别的植株数）/处理的总株数。

2.2.2 丙二醛含量的测定

参照 Dhindsa et al.（1981）的方法，略有改进。5%三氯乙酸提取，10 000 r/min 离心 15 min，加入等体积的反应液 0.5%硫代巴比妥酸，然后沸水浴 10 min，冷却之后进行吸光度的测定。

2.2.3 过氧化氢含量的测定

参照 Mukherjee and Choudhuri（1983）的方法，略有改进。1.0 g 叶片在液氮中迅速研磨成干粉，加－20℃预冷过的丙酮 2 mL 混匀。4℃下 12 000 r/min 离心 10 min。1 mL 上清液中加入 0.2 mL 氨水和 0.1 mL 浓盐酸（含 20%四氯化钛）。离心，预冷过的丙酮清洗生成的沉淀至无色，然后用 3 mL 2 mol/L 浓硫酸溶解沉淀，进行 410 nm 吸光度的测定。

2.2.4 抗坏血酸-谷胱甘肽循环代谢物的测定

叶片材料中加 50 mmol/L 磷酸缓冲液，pH 7.6（含 6.5 mmol/L EDTA，5 mmol/L AsA，4%PVP）进行提取。4℃下 13 000 r/min 离心 20 min。上清液用于酶活测定，所有操作 4℃进行。

APX 酶活测定参照 Nakano and Asada（1981）的方法。1.5 mL 反应体系：50 mmol/L 磷酸缓冲液（pH 7.0），0.1 mmol/L EDTA，0.5 mmol/L AsA，1.0 mmol/L H_2O_2 和 40 μL 酶液量。H_2O_2 启动反应。

MDHAR 酶活测定参照 Arrigoni et al.（1981）方法。1.0 mL 反应体系：50 mmol/L HEPES-KOH 缓冲液（pH 7.6），0.1 mmol/L NADH，2.5 mmol/L AsA，0.5 U AAO（抗坏血酸氧化酶）和 100 μL 酶液量，AAO 启动反应。

DHAR 酶活测定参照 Nakano and Asada（1981）方法。1.5 mL 的反应体系：50 mmol/L 磷酸缓冲液（pH 7.0），2.5 mmol/L GSH，0.1 mmol/L EDTA，0.2 mmol/L 脱氢抗坏血酸（DHA）和 60 μL 酶液量。DHA 启动反应。

GR 酶活测定参照 Schaedle（1977）方法。1.0 mL 反应体系：50 mmol/L 磷酸缓冲液（pH 7.6），0.4 mmol/L EDTA，0.2 mmol/L NADpH，0.5 mmol/L 氧化型谷胱甘肽（GSSG）和 75 μL 酶液量，NADpH 启动反应。

2.2.5 抗坏血酸和谷胱甘肽含量的测定

AsA 含量参照 Logan et al.（1998）方法测定。6% 高氯酸提取，4℃下 12 000 r/min 离心 20 min。100 μL 上清液中加入 20 μL 1.5 mol/L 碳酸钠中和 pH 为 1～2 200 mmol/L 醋酸缓冲液（pH 5.6）测定 265 nm 的吸光度。然后反应液里加入 1.5 U AAO，15 min 后再测吸光度。

还原型和氧化型谷胱甘肽测定参照 Griffith（1980）方法，略有改进。5% 磺基水杨酸提取，4℃下 12 000 r/min 离心 20 min。总谷胱甘肽含量测定反应体系：200 μL 0.5 mol/L 磷酸缓冲液（6.3 mmol/L EDTA，pH 7.5），560 μL 10 mmol/L EDTA，100 μL 6 mmol/L DTNB，100 μL 2.1 mmol/L NADpH 和 20 μL 上清液，1U GR 启动反应，测定 412 nm 下的吸光度。氧化型谷胱甘肽（GSSG）含量测定：20 μL 上清液加入 200 μL 磷酸缓冲液和 4 μL 2-乙烯吡啶（2-vinylpyridine），25℃孵育 30 min，然后参照总谷胱甘肽方法进行测定。还原型谷胱甘肽等于总谷胱甘肽减去氧化型谷胱甘肽。

2.2.6 抗氧化酶活测定

取材 1.0 g 放入预冷的研钵中，加入 5 mL 预冷的磷酸提取液进行研磨（100 mmol/L 磷酸缓冲液，pH 7.5，1 mmol/L EDTA 和 4%PVP）。4℃下 13 000 r/min 离心 20 min，上清液用于酶活测定。

总 SOD 酶活测定参照 Giannopolitis and Ries（1977）方法，略有改动。1 mL 反应体系：50 mmol/L 磷酸缓冲液 pH 7.8，6.5 mmol/L 蛋氨酸（MET），50 μmol/L NBT，20 μmol/L 核黄素（riboflavin），10 μmol/L EDTA 和 55 μL 酶液。不加酶液作为调零管。反应液避光混匀后放在 600 μmol/（m^2·s）光强下 5 min。测 560 nm 下的吸光度。一个单位 SOD 酶活定义为抑制 50%NBT 光还原所需要的酶量。

CAT 酶活表示过氧化氢分解引起的 240 nm 吸光度的下降［消光系数 39.4 L/（mmol/L·cm）］，参照 Aebi（1984）方法进行测定。1 mL 反应体系：50 mmol/L 磷酸缓冲液 pH 7.0，10 mmol/L H_2O_2（现配），和 50 μL 酶量。H_2O_2 启动反应，酶活单位为：μmol H_2O_2/(s·g)（以鲜重计）。

POD 酶活表示愈创木酚氧化引起的 470 nm 吸光度增加［消光系数为 26.8 L/（mmol/L·cm）］，按照 Hammerschmidt et al.（1982）方法进行测定。1 mL 反应体系：50 mmol/L 磷酸缓冲液 pH 7.0，10 mmol/L H_2O_2，10 mmol/L 愈创木酚和 50 μL 酶量。H_2O_2 启动反应，酶活单位：μmol H_2O_2/（s·g）（以鲜重计）。

2.2.7 RNA 提取和实时定量（qRT-PCR）分析基因表达

（1）取材：ABA 和清水喷施 72 h 后，对低温处理过的叶片进行取材，用 TRIZOL 试剂盒（Invitrogen，USA）按照说明书的操作步骤进行 RNA 提取。在 NanoDrop 仪器（Thermo Scientific NanoDrop 2000C Technologies，Wilmington，USA）进行 RNA 浓度的测定，A260/A280 和 A260/A230 的比值表示 RNA 的纯度。

①叶片用液氮迅速研成粉末，取 0.2 g 左右放到 2.0 mL 离心管，快速加 1.0 mL TRIZOL 提取液，混匀静置 5 min 使蛋白充分裂解。

②12 000 r/min 离心 10 min，4℃，取上清液。

③加入 0.5 mL 预冷过的氯仿，混匀静置 5 min 使水相与有机相充分分层。

④12 000 r/min 离心 10 min，4℃，取上清液中加入等体积的水饱和酚：氯仿（2：1），混匀。

⑤同样条件下离心，上清液再次加入等体积的氯仿进行抽提。

⑥离心后，小心吸取上清液，勿晃动分界层。

⑦加入等体积的异丙醇，-70℃沉淀 20 min。

⑧离心，倒掉上清液，离心管中加入 1 mL 75％无水乙醇冲洗两次。

⑨酒精挥发完，加入 20 μL ddH$_2$O（无 RNA 酶）溶解。

（2）按照 PrimeScript TM first-strand complementary DNA (cDNA) Synthesis Kit（TaKaRa，Japan）合成 cDNA。

①在 0.2 mL PCR 管中加入表 3-1 组分。

表 3-1 RNA 反转录反应体系

反应组分	加样量（μL）
5×PrimeScript™ 缓冲液（实时 PCR）	2
PrimeScript™ RT 酶混合物 I ATP（10 mmol/L）	0.5
Oligo dT 引物（50 μmol/L）	0.5
随机 6 聚体（100 μmol/L）	0.5
总 RNA（500 ng/μL）	1.0
无 Nase 水	5.5
总体积	10

②混匀后放在 PCR 仪 37℃延伸 15 min，85℃灭活 5 s。

（3）SYBR® Premix Ex Taq TM II（TaKaRa，Japan）进行实时定量分析

①在 iCycler iQ TM Multicolor PCR Detection System（Bio-Rad，Hercules，CA，USA）仪器上进行反应。反应体系见

表 3-2。

表 3-2　实时定量反应体系

反应组分	加样量（μL）
SYBR® Premix Ex Taq TM II	10
cDNA（50 ng/μL）	2.0
上游引物（10 mmol/L）	0.4
下游引物（10 mmol/L）	0.4
无菌水	7.2
总体积	20

②反应程序：95℃预变性 1 min，45 个循环：95℃ 变性 15 s，54℃ 退火 20 s，72℃延伸 30 s。在 54℃ 退火过程中采集荧光。

③辣椒中用微管蛋白和泛素连接蛋白基因（beta tubulin and ubiquitin-conjugating protein）作内参。试验中所用的基因引物序列见表 3-3。

表 3-3　基因引物序列

基因	登录号	上游和下游引物 $5'\rightarrow 3'$
POD	FJ596178.1	F：GCAGCATTCCTCCTCCTACT R：ATTTCTTTGCCTTGTTGTTG
Mn-SOD	AF036936.2	F：CTCTGCCATAGACACCAACTT R：CCAAGTTCGGTCCTTTAATAA
DHAR1	AY971873	F：TATCAATGGGCAGAATGTTT R：TCTCTTCAGCCTTGGTTTTC
DHAR2	AY971874	F：CCAAACCTCCGCTGACAAC R：AATCAGCAGCAGATACCTCAT
β-tubulin	EF495259.1	F：GAGGGTGAGTGAGCAGTTC R：CTTCATCGTCATCTGCTGTC
UBI-3	AY486137.1	F：TGTCCATCTGCTCTCTGTTG R：CACCCCAAGCACAATAAGAC

④基因的相对表达量按照 Delta-Delta Ct 方法进行计算。

2.2.8 数据统计分析

数值用平均值±误差表示。SAS 8.2 在最小显著差异水平进行方差分析，$P<0.05$ 认为差异显著。

2.3 结果与分析

2.3.1 ABA 对低温下辣椒叶片冷害指数的影响

外源 ABA 可有效地减轻低温下辣椒叶片的可见冷害症状（表 3-4）。低温 2 d 后，ABA 处理的植株表现出轻微的伤害（坏死斑面积小于 5%），而未处理的植株表现出严重的症状（坏死斑面积大于 25%）。喷施清水的幼苗表现出严重的萎蔫现象；而喷施 ABA 的叶尖少数略有卷曲，大多数叶片完全展开。

表 3-4 外源 ABA 对低温下辣椒叶片冷害症状的影响

处理	冷害指数
ABA+低温	1.73±0.20b
低温	3.50±0.26a

注：数值代表平均值±标准差，根据最小差异显著法分析不同的小写字母代表差异达显著水平（$P<0.05$）。

2.3.2 ABA 对低温下辣椒叶片中 MDA 和 H_2O_2 含量的影响

低温处理辣椒叶片，MDA（图 3-1A）和 H_2O_2（图 3-1B）含量显著增加（$P<0.05$），与对照相比，低温 6 h 分别增加 188% 和 122%。常温下 ABA 处理也增加 MDA（3 h 除外）和 H_2O_2（120 h 除外）含量。与低温处理相比，外源 ABA 降低遭受低温胁迫的叶片 MDA 和 H_2O_2 含量，3 h 达到显著水平（$P<0.05$）。总之，低温胁迫下辣椒叶片 MDA 和 H_2O_2 含量增加，而外源 ABA 处理明显降低了其含量。

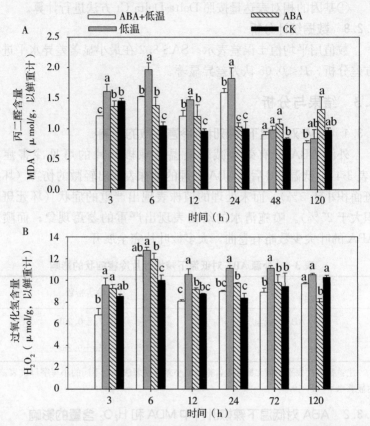

图 3-1 辣椒叶片中丙二醛 MDA (A) 和过氧化氢 H_2O_2 (B) 含量的变化

2.3.3 外源 ABA 对低温下辣椒叶片 AsA-GSH 循环代谢物的影响

低温处理期间谷胱甘肽还原酶 GR（图 3-2A）、脱氢抗坏血酸还原酶 DHAR（图 3-2B）和单脱氢抗坏血酸还原酶 MDHAR（图 3-2C）酶活变化趋势一致：12 h 或 24 h 期间酶活增加，随后下降，表明 AsA-GSH 循环在一定程度上缓解了低温伤害。与 CK（对照）相比，低温和 ABA 处理增加 GR、DHAR 和 MDAHR 酶活，低温 12 h 时分别比对照高出 28%、70% 和 37%。低温处理 GR 和 MDAHR 酶活高于 ABA＋低温处理，表明外源 ABA 和低温对其酶活的影响不是简单的协同关系。ABA＋低温处理的植株DHAR酶活12 h时显著低于低温处理组，24 h活性升高。

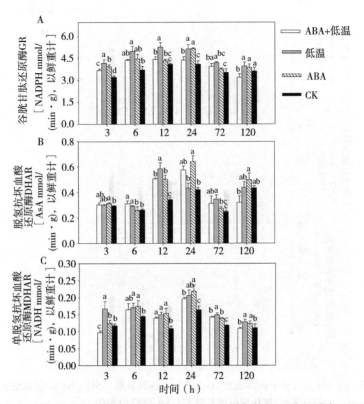

图3-2 辣椒叶片中谷胱甘肽还原酶GR（A），脱氢抗坏血酸还原酶DHAR（B）和单脱氢抗坏血酸还原酶MDHAR（C）酶活变化

与CK相比，低温处理叶片AsA和GSH含量增加，而ABA处理降低其含量（图3-3A和B）。ABA＋低温处理叶片AsA和GSH的增加水平低于低温处理组，表明其含量的增加是ABA和低温相互作用的结果。低温诱导AsA和GSH的含量分别是CK和ABA处理的180%（12 h）和170%（6 h）。低温72 h后，ABA或清水处理叶片GSH/GSSG比值明显增加（图3-3C），表明辣椒幼苗耐寒性逐渐增强（这与以上提到的MDA和H_2O_2含量的变化

趋势一致,图 3-1A 和 B)。

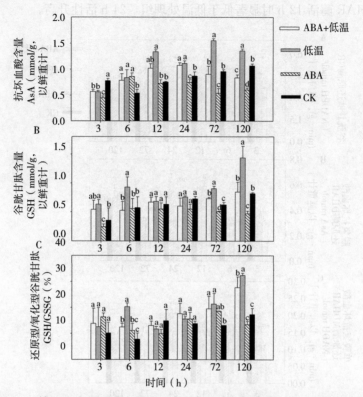

图 3-3 辣椒叶片中抗坏血酸 AsA（A）、谷胱甘肽 GSH（B）含量的变化和还原型/氧化型谷胱甘肽 GSH/GSSC 的变化

2.3.4 外源 ABA 对低温下辣椒叶片抗氧化酶的影响

三种处理叶片 SOD 酶活性 0~12 h 短暂下降,随后逐渐增加(图 3-4)。与 CK 相比,在低温处理 24 h 期间 SOD 酶活性下降,ABA 处理其酶活性增强。"ABA+低温"处理组 SOD 酶活性高于低温处理组,高 30%（3 h）;ABA 处理其酶活性高于低温处理,高 36%（3 h）。外源 ABA 提高常温和低温下辣椒幼苗 SOD 酶活性。

辣椒叶片中愈创木酚过氧化物酶 POD,过氧化氢酶 CAT 和抗坏血酸过氧化物酶 APX 酶活性变化如图 3-5。低温胁迫植株 POD 酶

第3章 辣椒对低温胁迫的响应与其低温抗性相关基因的克隆和功能分析

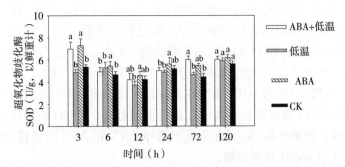

图3-4 辣椒叶片超氧化物歧化酶SOD酶活性的变化

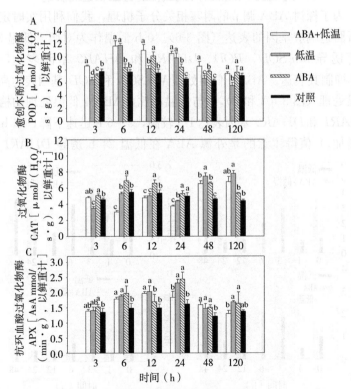

图3-5 辣椒叶片中愈创木酚过氧化物酶POD（A），过氧化氢酶CAT（B）和抗坏血酸过氧化物酶APX（C）酶活性变化

活性 6 h 达到峰值而后大幅度下降。与 CK 相比，POD 在 6～48 h 低温处理期间活性增加，CAT 酶活性在低温 48 h 后增加。外源 ABA 显著增加 POD 和 CAT 酶活性（$P<0.05$）。与低温处理相比，外源 ABA 增加低温胁迫下叶片 POD 酶活性（6 h、24 h 除外），降低 CAT 和 APX 酶活性。而且，12 h 和 48 h 的 ABA+低温处理 POD 酶活性显著高于仅进行低温处理的 POD 酶活性。因此，我们推测外源 ABA 在常温下诱导 CAT 和 APX 酶活性增加，而在低温下不诱导其增加。

2.3.5 ABA 对低温下辣椒叶片抗氧化酶基因表达的影响

为了探讨 ABA 调节的耐寒相关分子机理，我们利用实时定量分析胁迫响应基因的表达（图 3-6）。0 h 低温作为对照。低温 3 h 显著诱导 $Mn\text{-}SOD$、POD、$DHAR1$ 和 $DHAR2$ 基因表达，而 12 h 抑制其表达。ABA+低温处理 $Mn\text{-}SOD$ 和 POD 基因表达高于低温处理（图 3-6A 和 B）。（与低温相比，ABA+低温处理的植株 $DHAR1$ 和 $DHAR2$ 基因表达 1 h 较高，3～6 h 迅速下降，24 h 显著增加。）值得注意的是外源 ABA 在低温 24 h 诱导 $DHAR1$ 和

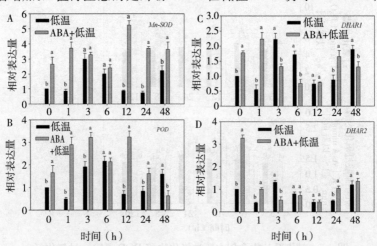

图 3-6 辣椒叶片中 $Mn\text{-}SOD$（A），POD（B），$DHAR1$（C）和 $DHAR2$（D）基因表达的变化

DHAR2 基因表达。ABA+低温处理 *Mn-SOD*、*POD*、*DHAR1* 和 *DHAR2* 的最高基因表达量分别高于低温处理 5.01、3.41、3.07 和 2.27 倍。

2.4 讨论

2.4.1 ABA 对低温下辣椒叶片冷害症状的影响

ABA 处理植株表现较小的坏死斑面积，未处理植株冷害较重，出现严重的萎蔫。这些结果表明 ABA 对辣椒幼苗冷害的保护作用，可有效地预防低温胁迫下叶片失水。

2.4.2 ABA 对低温下辣椒叶片 MDA 和 H_2O_2 含量的影响

冷害引起失水，最终导致萎蔫和膜完整性破坏。MDA 是衡量胞膜损坏程度的指标，因此与冷害指数有关。低温增加 MDA 和 H_2O_2 含量（Li et al., 2011），ABA 处理降低其含量（Nayyar et al., 2005）。在本研究中，低温显著增加辣椒幼苗 MDA 和 H_2O_2 含量，表明抗氧化酶不足以清除过量活性氧的积累。喷施外源 ABA 的植株在低温下其含量显著降低，表明外源 ABA 通过降低 MDA 和 H_2O_2 含量在逆境胁迫下起到保护作用。

2.4.3 外源 ABA 对低温下辣椒叶片抗氧化酶活性的影响

非生物胁迫引起活性氧增加，植株表现的耐逆性与小分子抗氧化物和清除活性氧相关的酶有关。除了 CAT、SOD 和 POD 外，AsA-GSH 循环直接参与活性氧清除的防御反应。目前研究表明 ABA 通过提高作物的抗氧化能力避免非生物胁迫（Ding et al., 2010；Wang et al., 2011）。本试验主要探讨外源 ABA 对 AsA-GSH 循环代谢物（APX、DHAR、MDHAR 和 GR）的影响。

植物 MDHAR 和 DHAR 通过再生 AsA，在抗氧化反应中起着重要作用。GR 是 AsA-GSH 循环途径中重要的抗氧化酶。在 NADpH 作用下，还原氧化型谷胱甘肽（GSSG）生产还原型谷胱甘肽（GSH）。在本研究中，低温引起辣椒 GR、DHAR 和 MDHAR 活性增加，表明 AsA 充分再生进行活性氧的清除（图 3-2），也被其他的研究结果证实（Zhang et al., 2009）。然而，与低温处理相比，ABA 预处理遭受低温的植株以上提及的酶活性降低。这种结果可能与 ABA＋低温处理氧化胁迫程度不足以启动 AsA-GSH 循环有关（ABA 对低温下辣椒冷害指数、MDA 和 H_2O_2 含量的影响支持这一假设，表 3-1 和图 3-1）。ABA＋低温处理 DHAR 活性在 24 h 高于低温处理，与下面描述的结果一致：在 24 h 时，ABA 预处理植株 AsA 含量与低温处理植株的含量一样。这些结果表明外源 ABA 主要通过 DHAR 酶提高低温下辣椒叶片 AsA 含量。

研究表明 AsA 和 GSH 在非生物胁迫的抗性中起着重要作用。比如，低温增加黄瓜叶片 AsA 和 GSH 含量（Li et al.，2011）。拟南芥 AsA-GSH 循环再生 AsA 含量的增加，减弱逆境诱导的氧化胁迫（Wang et al.，2010）。在低温胁迫下辣椒叶片 AsA 和 GSH 含量增加（图 3-3），归因于 MDHAR、DHAR 和 GR 活性增强。考虑到 AsA 和 GSH 含量的增加与 H_2O_2 含量的降低趋势一致，表明 AsA-GSH 循环参与低温下辣椒活性氧的清除。相反，与低温处理相比，ABA 预处理降低低温胁迫下辣椒 AsA 和 GSH 含量，与 AsA-GSH 循环中的代谢酶（APX、GR、DHAR 和 MDHAR）变化一致。这些结果表明在低温胁迫中，外源 ABA 没有诱导辣椒叶片 AsA-GSH 循环途径进行活性氧的清除。此结果不同于以前的报道，如其他的外源物质增加非生物胁迫下 AsA 和 GSH 含量及其相关酶活（Liu et al.，2010；Shan，Liang，2010）。不同的结果或许与植物种类、胁迫程度和时间及外源物质有关。然而，外源 ABA 在缓解低温下辣椒冷害指数，MDA 和 H_2O_2 水平方面起到保护作用（表 3-1 和图 3-1），我们推测在 ABA 预处理遭受低温胁迫的植物中，其他的抗氧化酶负责进行过量活性氧的清除。

在各种细胞中 SOD 高效地歧化超氧阴离子（O_2^-）生成 H_2O_2。低温显著降低黄瓜 SOD 酶活（Zhang et al.，2009）。ABA 诱导干旱胁迫下 SOD 酶活（Wang et al.，2011）。本试验中低温 24 h 期间 SOD 活性下降（图 3-4），导致 H_2O_2 和 MDA 过量积累（图 3-1）。与低温处理相比，外源 ABA 增强低温下辣椒叶片 SOD 酶活，表明喷施 ABA 的植株在低温下清除 O^{-2} 的能力较强。耐寒性直接与 SOD 酶活有关（de Azevedo Neto et al.，2005）。

SOD 催化 O^{-2} 生成 H_2O_2 进而缓解植株的毒害作用，但过量的 H_2O_2 积累对植株的生长也是有毒害作用。在植物中，很多酶与 H_2O_2 代谢有关，主要有 CAT、APX 和 POD。低温增加黄瓜叶片 APX、GR 和 POD 酶活，降低 CAT 活性（Lee and Lee，2000）。ABA 预处理诱导非生物逆境胁迫 SOD 和 POD 活性增加（Wang et al.，2011）。辣椒叶片 POD（图 3-5A）和 APX 活性（图 3-5C）在低温下增强，CAT 活性在低温 24 h 前降低（图 3-5B），表明清除活性氧的保护机制主要依靠 POD 和 APX，随着处理时间的延长，协同 SOD 和 CAT 的作用。然而，低温处理植株 H_2O_2 含量最高，表明 H_2O_2 的产生超过了细胞内抗氧化酶的清除能力。CAT 活性下降可能由于低温诱导的 H_2O_2 积累没有启动 CAT；考虑到低温显著诱导辣椒 AsA-GSH 循环，我们推测当 CAT 活性下降时，这一循环主要负责清除过量的 H_2O_2。这种假设也被其他的研究所证实（Lee and Lee，2000）。与单一的低温处理相比，ABA 喷施的植株在低温下 SOD 和 POD 活性增强，CAT 和 APX 活性下降。这些结果表明在遭受低温胁迫的辣椒叶片，外源 ABA 主要通过 SOD 和 POD 增强清除过量 O^{-2} 和 H_2O_2 的能力。

ABA 在提高耐寒性和促进衰老方面起着重要作用。ABA 信号途径参与低温和衰老调控，由信号分子蔗糖和活性氧介导。ABA 诱导水稻衰老叶片 MDA 和 H_2O_2 积累，加快叶片衰老（Hung and Kao，2004）。ABA 也能提高抗氧化酶活性比如 SOD、APX 和 GR，进而减缓叶片衰老（Hung and Kao，2003）。在本试验中，常温下喷施 ABA 显著增加 MDA 和 H_2O_2 含量（图 3-1），及 GR、DHAR、MDHAR、SOD、POD、CAT 和 APX 酶活（图 3-2，3-4，3-5）；降低 AsA 和 GSH 含量（图 3-3）。这些结果与以上提到的 ABA 诱导衰老相似（Hung et al.，2003，2004）。

2.4.4　ABA 对低温下辣椒抗氧化酶基因表达的影响

Mn-SOD 基因强烈响应低温和氧化胁迫反应（Lee and Lee，

2000；Li et al.，2009），ABA 诱导其表达（Bueno et al.，1998）。本试验中，ABA 通过诱导基因表达提高耐寒性，如参与氧化胁迫防御机制的 *Mn-SOD* 和 *POD*。值得注意的是，ABA 显著提高低温 24 h 时辣椒 *DHAR1* 和 *DHAR2* 基因表达，与 DHAR 酶活变化一致（图 3-2B）。总之，ABA 介导的 SOD、POD 和 DHAR 酶活变化受相应基因的转录水平调控。ABA 诱导的抗氧化酶基因调节低温引起的活性氧含量，有助于提高耐寒性（Xue-Xuan et al.，2010）。

结论：本研究揭示 ABA 在调节低温反应的生理和生化代谢。喷施外源 ABA 是提高辣椒耐寒性的有效方法。这种结果部分归功于 SOD 和 POD 酶活及其基因表达的显著增加，也与低温下辣椒 AsA-GSH 循环密切相关。本研究发现 ABA 的保护机制有助于提高植物的抗逆性。然而，ABA 诱导的耐寒性机理和叶片衰老之间的关系有待于进一步研究。

3 SSH 技术对 ABA 调节的辣椒低温抗性相关基因的差异表达分析

为了适应低温胁迫，植物体内发生一系列的基因表达、生理生化和形态变化，进而提高植物的耐寒性（Smallwood et al.，2002；Zhu et al.，2007）。外源 ABA 模拟逆境胁迫，从而在这一变化过程中起着重要的作用（Bray，1988；Loik et al.，1993）。目前，关于 ABA 调节基因表达的研究越来越多。ABA 主要在转录水平调节拟南芥基因表达，已经鉴定 1 350 多个基因在响应 ABA 信号转导中受其调节表达（Chak et al.，2000；Seki et al.，2002）。采用微芯片技术分离 ABA 响应的转录组报道很多，如拟南芥和水稻（Matsui et al.，2008；Rabbani et al.，2003）、高粱（Buchanan et al.，2005）和其他的植物种类（Vij et al.，2007）。虽然关于以上提及的植物中 ABA 调节的转录组信息很多，但在其他的冷敏感植物中，对 ABA 调节的逆境相关基因直接进行研究还是很有必要的。Guo et al.（2012）报道外源 ABA 能够提高辣椒对低温诱导的抗氧化胁迫能力，主要归因于抗氧化酶活及其相关基因表达的提高。然而在辣椒上响应低温胁迫反应时，ABA 介导的基因调节还没有完全分离和鉴定。关于辣椒逆境响应因子信息的增加有助于增强辣椒的抗逆性，进而提高辣椒产量。

抑制消减杂交文库（SSH）是一种分离和鉴定低丰度差异基因的有效方法（Diatchenko et al.，1996）。与其他的技术相比如 DNA 微芯片技术，SSH 主要的优点在于均一化和杂交结合，分离高丰度的差异基因和富集低丰度的转录（Clement et al.，2008）。而且，它不需要全部的基因组信息，能够克服全部基因表达分析的限制。因此，SSH 可以有效研究缺少大量序列信息的植物（Gulyani et al.，2011）。目前，SSH 技术已经被用来研究植物响应逆境反应，以期提高其抗逆性（Nguyen et al.，2009；Gulyani et al.，2011）。

本试验，叶面喷施 ABA 和清水，72 h 后进行低温 48 h 处理，对不同低温时间段的叶片进行 RNA 提取，用来构建正反交 SSH 文库。通过核酸序列的同源性分析 ABA 上下调控的基因，并对这些基因的表达进行实时定量分析。分离和鉴定 ABA 调节的抗寒相关基因可以为提高植物的抗逆性提供有效的信息。

3.1 材料与处理

辣椒品种 P70 种子催芽后播种在营养钵中，按照以前描述的方法放置在光照培养箱进行培养（Guo et al.，2012）。植株生长条件设置为：温度 25/18℃（白天/黑夜），12 h 光周期，光强 20 000 lx，相对湿度 60%～70%。当植株生长到 6～8 片真叶，进行 3 种处理：对照，喷施蒸馏水；2 个 ABA 处理，早上 9：00—10：00 喷施 0.57 mmol/L ABA 至布满叶的上下表面，叶面喷施 72 h 后进行低温 6℃处理；另一组 ABA 处理放置在常温下，12 h 光周期，光强 5 500 lx。ABA 处理和对照的植株叶片在不同的低温处理时间段进行取样：0、1、3、6、12、24 和 48 h。每个时间点，从 4 株植株上取 2 或 3 片上部叶片用锡箔纸包成一小包,迅速放入液氮中,-80℃储存。处理随机排布,4 次重复。

3.2 试验方法

3.2.1 叶绿素含量测定

总叶绿素含量测定参照 Korkmaz et al.（2010）的方法。0.5 g 叶片用 5 mL 80% 丙酮（体积比）研磨。按照 Lichtenthaler（1987）提出的公式 [Chl a+b (mg/g FW) = $7.79 \times A 663 + 16.26 \times A 645$] 进行计算 3 次重复,每次重复随机取材于 3 株植物的新鲜叶片。

3.2.2 电导率测定

电导率测定参照 Dionisio-Sese and Tobita（1998）的方法。10 个叶圆片（直径 1 cm）随机取样于 2 株植株的上部,完全展开叶片。蒸馏水清洗叶表面污染物。加入 10 mL 蒸馏水,常温下放置 2 h。电导仪（DDS-307,中国）测起始电导率 $EC1$。然后,样品沸水浴 30 min 释放离子,待冷却至常温后,测最终电导率 $EC2$。电导率按公式计算 $EC1/EC2$,百分数表示。

3.2.3 净光合速率和气孔导度分析

净光合速率 P_n 和气孔导度 G_s 用便携式光合仪 LI-6400XT（Li-Cor Inc.，Lincoln，NE，USA）测定。叶片取自植株的相同叶位（从顶端起第 3 或 4 片真叶）,5 次重复。叶室条件设为:6℃,70% 相对湿度,PAR 100 $\mu mol/L/(m^2 \cdot s)$ 和 CO_2 浓度 520 $\mu mol/L$。

3.2.4 RNA 分离和 cDNA 准备

TRIZOL 试剂盒（Invitrogen Carlsbad，CA，USA）根据说明书上的步骤提取 RNA。SSH 文库构建,ABA 处理的材料总 RNA 由低温下各个时间点的 RNA 等量混合而成（0、1、3、6、12、24 和 48 h）。类似地,蒸馏水喷施的材料总 RNA 由低温下相同时间点的 RNA 等量混合。PolyATtract® mRNA Isolation Systems Ⅳ（Promega，USA）试剂盒按照说明书的步骤分离 Poly

(A)＋mRNA。NanoDrop 仪器测定总 RNA 和 mRNA 浓度；A260/280 和 A260/230 比值和电泳检测其纯度。

3.2.5 SSH 文库构建

正反交 SSH 文库参考 PCR select-cDNA SSH kit（Clontech，Palo Alto，USA）试剂盒进行构建。ABA 和蒸馏水（对照）预处理的低温植株 cDNA 进行 RsaI 酶切 37℃，1.5 h，酚/氯仿抽提，乙醇沉淀，最后重悬于无菌水。RsaI 酶切后的 cDNA（清水处理对照 C 和 ABA 处理 T）分成 4 份。其中一份分别进行 16℃过夜连接接头-1，连接产物分别为 CA1（RsaI 酶切后 cDNA 带有接头-1 来自蒸馏水预处理的低温植株材料）和 TA1（RsaI 酶切后 cDNA 带有接头-1 来自 ABA 预处理的低温植株材料）；另外一份按照相同的方法连接接头-2R，分别称为 C2R 和 T2R。剩下的两份代表 RsaI 酶切后的 cDNA 末端，被称为 driver。

正交 SSH 文库，代表过量和诱导的转录基因，TA1 和 T2R 分别称为 tester-a 和 tester-b；C 处理作为 driver。反交 SSH 文库按照相反的方法构建，代表下调转录基因的富集。比如，CA1 和 C2R 分别作为 tester-a 和 tester-b；T 作为 driver。进行二次杂交，driver cDNA 加入每管 tester 中，随后重悬于杂交缓冲液。热变性后，混合物 68℃退火 8 h。第 1 次杂交后混样，新变性 cDNA 再次加入进行第 2 次杂交。2 次杂交后的产物进行 PCR 扩增，产物直接连入 pGEM-T easy 载体和转化大肠杆菌 DH5α。重组子进行蓝白斑筛选。

3.3 结果与分析

3.3.1 ABA 有效地缓解辣椒叶片的冷害症状

外源 ABA 有效地缓解低温下辣椒叶片的冷害症状（图 3-7）。未喷施 ABA 的材料低温 1 h 出现萎蔫；而喷施 ABA 的植株直到低温 12 h 才出现萎蔫。低温 6 h 对照材料茎下垂，24 h 后恢复直立，48 h 后叶片逐渐展开，这些结果表明辣椒对低温的适应反应。

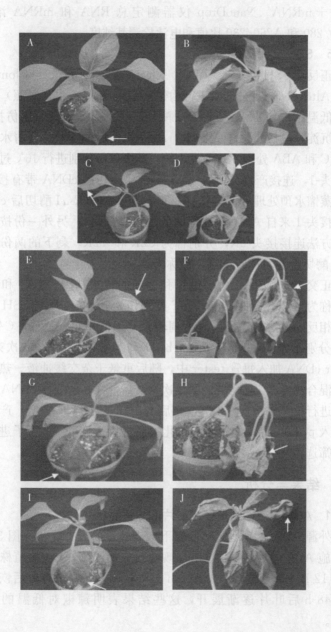

图 3-7 外源 ABA 对辣椒幼苗冷害症状的影响

A. 0.57 mM ABA 预处理+低温胁迫 1 h；B. 清水预处理+低温胁迫 1 h；C. 0.57 mM ABA 预处理+低温胁迫 3 h；D. 清水预处理+低温胁迫 3 h；E. 0.57 mM ABA 预处理+低温胁迫 6 h；F. 清水预处理+低温胁迫 6 h；G. 0.57 mM ABA 预处理+低温胁迫 12 h；H. 清水预处理+低温胁迫 12 h；I. 0.57 mM ABA 预处理+低温胁迫 24 h；J. 清水预处理+低温胁迫 24 h；K. 0.57 mM ABA 预处理+低温胁迫 48 h；L. 清水预处理+低温胁迫 48 h

3.3.2 ABA 对低温下辣椒电导率、叶绿素、净光合速率和气孔导度的影响

经低温处理的植株电导率增加（图 3-8A）。低温诱导的植株电导率高于 ABA 处理的植株电导率。低温引起叶绿素在未处理植株的下降量高于 ABA 处理的植株（图 3-8B）。净光合速率在整个低温处理

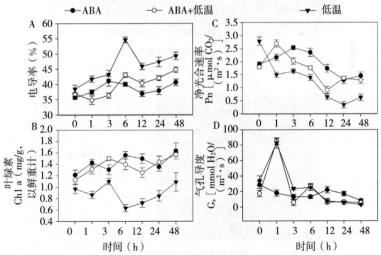

图 3-8 外源 ABA 对辣椒叶片中电导率、叶绿素、净光合速率和气孔导度的影响

期间迅速下降（图 3-8C）。然而，外源 ABA 可有效地减缓低温引起的净光合速率下降。ABA＋低温处理和低温处理的植株气孔导度在 1 h 后快速达到峰值，之后 2 h 又迅速下降（图 3-8D）。与对照相比，ABA 处理气孔导度在 3 h 降低，随后变化基本一致，表明低温 6 h 后，ABA 处理对辣椒叶片的气孔运动没有影响。总之，外源 ABA 减缓低温引起的电导率增加，叶绿素 a 损失和净光合速率下降，进而提高辣椒的耐寒性。

3.3.3 消减 cDNA 文库的构建

为了分离低温下 ABA 预处理的辣椒幼苗差异基因，本试验构建了正、反交文库。总 RNA 的检测标准在材料和方法中描述（图 3-9）。分离的 mRNA 大小主要位于 250～2 000 bp（图 3-10）。为了检测消减效率，二次 PCR 产物在 1‰电泳图所示（图 3-11）。PCR 片段主要介于 300～750 bp（图 3-12）。这些结果表明消减 cDNA 文库的高质量。

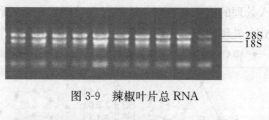

图 3-9　辣椒叶片总 RNA

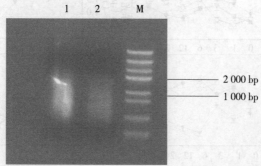

图 3-10　遭受低温胁迫的辣椒幼苗 mRNA

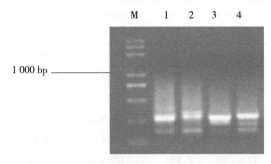

图 3-11　正反交文库结果，二次 PCR 产物电泳图

泳道 1：反交-未消减 cDNA；泳道 2：正交-未消减 cDNA；泳道 3：反交-消减 cDNA；泳道 4：正交-消减 cDNA

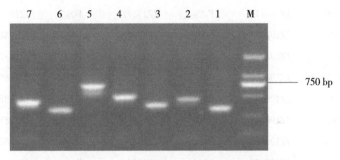

图 3-12　PCR 扩增出的杂交 cDNA 片段

3.3.4　正反交文库中 ESTs 分离

分别挑选正交库中 438 克隆点和反交库中 330 克隆点 PCR 扩增检测后，进行 DNA 测序。除去 ESTs 序列上的载体和接头，以及小于 100 bp 的片段，正反交文库分别获得 126 和 109 个高质量的 ESTs。CAP3 拼接软件分析序列的冗余性。235 个高质量的 ESTs 拼接成 18 个 contigs 和 55 个 singlets，总共 73 个 unigenes。其中，40 个 unigenes 在正交库中受 ABA 上调表达；在反交库中获得 33 个 unigenes，受 ABA 下调表达。这些 unigenes 提交到 GenBank dbEST，登录号为 JZ198744-JZ198816（表 3-5）。

表 3-5 在低温胁迫下 ABA 调节的辣椒 unigenes 登录号

Clone ID (编号)	Accession no. (登录号)	Clone ID (编号)	Accession no. (登录号)	Clone ID (编号)	Accession no. (登录号)
F001	JZ198744	F002	JZ198745	F003	JZ198746
F004	JZ198747	F005	JZ198748	F006	JZ198749
F007	JZ198750	F008	JZ198751	F009	JZ198752
F010	JZ198753	F011	JZ198754	F012	JZ198755
F013	JZ198756	F014	JZ198757	F015	JZ198758
F016	JZ198759	F017	JZ198760	F018	JZ198761
F019	JZ198762	F020	JZ198763	F021	JZ198764
F022	JZ198765	F023	JZ198766	F024	JZ198767
F025	JZ198768	F026	JZ198769	F027	JZ198770
F028	JZ198771	F029	JZ198772	F030	JZ198773
F031	JZ198774	F032	JZ198775	F033	JZ198776
F034	JZ198777	F035	JZ198778	F036	JZ198779
F037	JZ198780	F038	JZ198781	F039	JZ198782
F040	JZ198783	R001	JZ198784	R002	JZ198785
R003	JZ198786	R004	JZ198787	R005	JZ198788
R006	JZ198789	R007	JZ198790	R008	JZ198791
R009	JZ198792	R010	JZ198793	R011	JZ198794
R012	JZ198795	R013	JZ198796	R014	JZ198797
R015	JZ198798	R016	JZ198799	R017	JZ198800
R018	JZ198801	R019	JZ198802	R020	JZ198803
R021	JZ198804	R022	JZ198805	R023	JZ198806
R024	JZ198807	R025	JZ198808	R026	JZ198809
R027	JZ198810	R028	JZ198811	R029	JZ198812
R030	JZ198813	R031	JZ198814	R032	JZ198815
R033	JZ198816				

对正反交文库中的 unigenes 进行 blastx 和 blastn 分析，并利

第3章 辣椒对低温胁迫的响应与其低温抗性相关基因的克隆和功能分析

用 GO 软件注解其功能。正交库中 10 个 unigenes（25%）和反交库 5 个 unigenes（15.15%）在 NCBI 数据库中没有同源序列。正交库中 11 个 unigenes（27.5%）和反交库 10 个 unigenes（30.30%）与未知功能的序列同源。正交库中 19 个 unigenes 与已知功能的基因显著同源，并对其进行 GO 生物学功能注解。19 个 unigenes 分成 8 类基本功能，包括转录、信号转导、防御反应、运输和能量代谢（图 3-13）。参与转录调节的基因占 12.5%，能量代谢占 10%，防御反应占 7.5%，信号转导占 5% 和 ATP 代谢占 5%。反交库中 18 个 unigenes 与已知功能的基因显著同源，主要参与 7 类基本功能（图 3-14），分别为防御反应（18.18%）、脂肪、萜类化合物和氨基

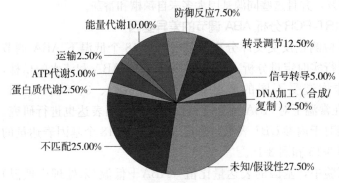

图 3-13 正交库中 unigenes 的功能分类

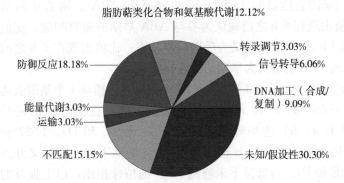

图 3-14 反交库中 unigenes 的功能分类

酸代谢 (12.12%)、DNA 加工处理 (9.09%)、信号转导 (6.06%)、转录 (3.03%)、能量代谢 (3.03%) 和运输 (3.03%)。

3.3.5 参与耐寒性的基因

进一步分析低温下辣椒响应 ABA 的分子机制，本试验从正反交文库中分离可能参与 ABA 调节的耐寒相关基因，并对其功能分成 6 类，如能量代谢、防御反应、信号转导和转录。这些基因如表 3-5 所示，包括乙醇脱氢酶、超氧化物歧化酶、质膜钙离子 ATP 酶和 F-盒蛋白（Guo et al.，2012；Schiøtt et al.，2005；Seki et al.，2002；Zhang et al.，2008），冗余率分别为 0.79% 和 0.92%；编码 NAC 转录因子的基因表现 9.52% 冗余率；与未知功能基因同源为 10%~25%，并且这些同源基因大多来自辣椒和番茄。

3.3.6 qRT-PCR 分析 ABA 调节的差异基因

为了检测正反交文库分离的基因，对 18 个低温下 ABA 调节的基因进行实时定量分析。总 RNA 来自喷施 ABA 和清水的植株，这些植株均遭受不同的低温处理时间（0、1、3、6、12、24、48 h）。在常温下对 ABA 预处理植株相应的基因表达也进行研究。表达量相对于内参 UBI 表示，来自表 3-6 中的 18 个基因表达量的变化如图 3-15 和图 3-16。

本试验中，基因的表达量比值（ABA＋低温/未处理＋低温）在任意一个时间点大于 3 被定义为 ABA 上调表达的基因；类似地，相应基因表达量的比值小于 1/3 被认为是 ABA 下调表达的基因。比值出现在 6 h 之前被认为参与 ABA 早期的短暂响应；比值出现在 24 h 或 48 h 被定义为后期响应基因；比值出现在 6 h 之前和 24 h 或 48 h 被定义为早期的持续响应基因。

在正交库中，低温 1 h 和 3 h ABA 预处理的植株 4 个基因表达（F007、F026、F030 和 F031）迅速增加，随着低温时间的延长（12、24 和 48 h）达到峰值（图 3-15A、B、C 和 D）。F027 和 F039 基因表达在低温 1 h 达到峰值，6 h 略微下降，随后又升高（图 3-15E 和 F）。与低温下未喷施 ABA 的植株相比，以上提及的 6 个基因表达量在 1 或 6 h 和 24 或 48 h 增加 3 倍，表明这些基因是

第3章 辣椒对低温胁迫的响应与其低温抗性相关基因的克隆和功能分析

表 3-6 在低温适应过程中参与 ABA 调节的 unigenes

基因编号	长度 (bp)	登录号	同源基因	同源基因登录号	同源率 (%)[a]	E 值	EST 重复率 (%)[b]
Functional category-transcription							
F007[c]	326	JZ198750	nam-like protein 4 (*Petunia ×hybrida*)	AF509867.1	82	1e-28	9.52
F027[c]	292	JZ198770	MADS-box transcription factor (*C. annuum*)	DQ999998.1	85	1e-51	0.79
R028[c]	332	JZ198811	Multiprotein bridging factor 1 (*S. tuberosum*)	EU294363.1	80	1e-47	0.92
Functional category-energy metabolism							
F026[c]	295	JZ198769	carbonic anhydrase (*S. lycopersicum*)	NM001247119.1	91	2e-100	0.79
F035[c]	309	JZ198778	alcohol dehydrogenase (*N. tabacum*)	X81853.1	86	3e-91	0.79
F039[c]	445	JZ198782	ribulose-1, 5-bispH ospH ate carboxylase/ oxygenase small subunit (*C. annuum*)	AF065615.1	99	8e-165	0.79
R017	298	JZ198800	cytochrome P450 like_ TBP (*N. tabacum*)	D64052.1	98	5e-51	0.92
Functional category-metabolism							
R011[c]	236	JZ198794	glyceraldehyde-3-pH ospH ate dehydrogenase (*C. annuum*)	AJ246011.1	82	4e-25	0.92
R032[c]	296	JZ198815	o-dipH enol-O-methyltransferase (*N. tabacum*)	X71430.1	92	3e-116	0.92
R029[c]	262	JZ198812	F-box protein (*Medicago truncatula*)	XM_ 003592058.1	75	1e-13	0.92
F038[c]	1110	JZ198781	Elongation Factor 2 (*Hordeum vulgare*)	AK362029.1	73	9e-74	0.79

(续)

基因编号	长度(bp)	登录号	同源基因	同源基因登录号	同源率(%)[a]	E值	EST重复率(%)[b]
Functional category-defense response							
F008[c]	311	JZI98751	chitin binding protein (*C. annuum*)	AF333790.1	90	1e-96	0.79
F031[c]	362	JZI98774	superoxidase dismutase (*S. lycopersicum*)	NM_001247840.1	78	2e-75	0.79
R031[c]	261	JZI98814	dehydrin-like protein (*C. annuum*)	AY225438.1	84	4e-51	0.92
Functional category-signal transduction							
F037[c]	273	JZI98780	serine/threonine kinase (*N. tabacum*)	DQ459385.1	85	6e-24	0.79
R030[c]	210	JZI98813	CBL-interacting serine/threonine-protein kinase 1 (CIPK1) (*A. thaliana*)	NM_202599.1	76	4e-06	0.92
Functional category-transport							
F030[c]	192	JZI98780	plasma membrane calcium ATPase (*A. thaliana*)	NM119927.2	76	1e-17	0.79
Functional category-unknown function							
F012[c]	219	JZI98755	*Capsicum annuum*	GU048903.1	83	5e-24	18.2
F029[c]	343	JZI98772	*Solanum lycopersicum*	AK322634.1	73	2e-43	10.3
R002[c]	1072	JZI98785	*Capsicum annuum*	GU048902.1	81	3e-54	10.09
R004[c]	513	JZI98787	*Capsicum annuum*	JF330775.1	71	4e-04	25.68

备注：利用CAP3拼接软件把正反库中ESTs拼接成singletons和contigs，然后在NCBI数据库中进行同源分析。正交库中unigenes命名为'F00X'，反交库中unigenes命名为'R00X'。a代表同源率；b代表一个unigene的ESTs数目/总ESTs；c代表进一步实时定量分析的克隆点。

第3章 辣椒对低温胁迫的响应与其低温抗性相关基因的克隆和功能分析

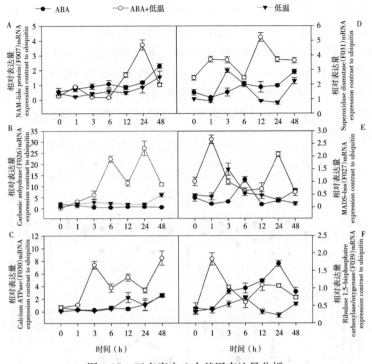

图 3-15 正交库中 6 个基因表达量分析

早期的持续上调基因。其中，F007 和 F027 分别与 NAC 和 MADS-box 转录因子高度同源，参与转录调节；F026（carbonic anhydrase），F030（calcium ATPase）和 F039（RuBisCo）参与代谢调节；F031（SOD）参与防御响应。另外，参与早期短暂响应基因（F008 和 F012）（图 3-16A 和 B）和后期上调响应基因（F029 和 F038）（图 3-16C 和 D）表达量比值达到 3 倍分别发生在 6 h 或 1 h 和 24 h。然而，参与代谢调节的延伸因子（F038）表达比值在 24 h 高于 2 倍。在整个低温处理期间，来自 ABA 处理以上提到的基因表达量高于对照，表明在低温下辣椒这些基因受 ABA 正调节。同时，常温下 ABA 预处理的植株 6 个基因表达量（F026、F030、F027、F008、F029 和 F038）（图 3-15B、C、E 和图 3-15A、C 和 D）基本没有变化，低于 ABA＋低温处理，表明 ABA 诱导的基因表达依靠低温。常温下 ABA 处理的植株 F039 表达（图 3-15F）逐渐增加，高于 ABA+低温处理。然而，低温处理 F039 表达基本没有变化，表明 ABA 上调此基因表达不依靠低温。来自 ABA 或低温单因素处理的 F007 表达在整个处理期间逐渐增加（图 3-15A）。对照 F030 表达在低温 12 h 后逐渐增加（图 3-15C）。F038 表达在低温 1 h 迅速增加，保持在高水平（6 h 除外），48 h 下降到起始水平（图 3-16D）。

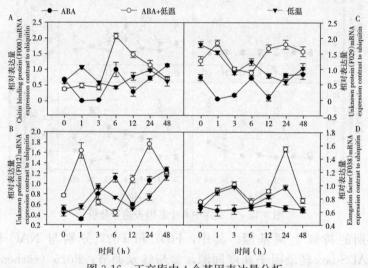

图 3-16　正交库中 4 个基因表达量分析

在反交库中，未处理植株 4 个基因（R011、R028、R029 和 R030）（图 3-17A、B、C 和 D）在低温 3 或 6 h 达到峰值，24 h 下降，48 h 又升高。参与防御反应的脱水素基因（R031）逐渐增加，48 h 上升到峰值（图 3-18D）。然而，与仅进行低温处理的植株相比，ABA+低温处理以上提及的基因表达量下降，且比值（ABA+低温/未处理+低温）低于 1/3，表明在低温下这些基因受 ABA 影响下调表达。R028、R029 和 R004 表达比值（图 3-17B、C 和图 3-18B）小于 1/3 发生在 3 h 和 48 h，属于早期的持续下调基因。R011、R030、R032（图 3-17A、D 和图 3-18C）和 R002、R031（图 3-18A、D）表

第3章 辣椒对低温胁迫的响应与其低温抗性相关基因的克隆和功能分析

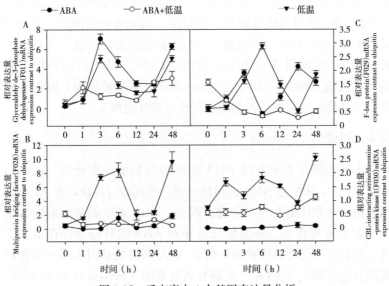

图3-17 反交库中4个基因表达量分析

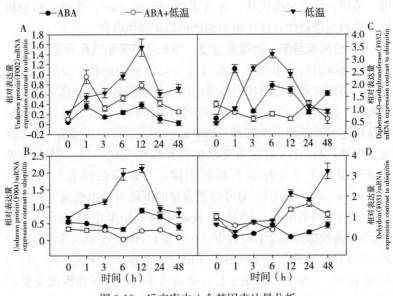

图3-18 反交库中4个基因表达量分析

达比值分别发生在 1 h 或 3 h 和 48 h，分别属于早期短暂和后期下调基因。R028 参与转录调节，与 MBF（转录辅激活因子）高度同源；R030 与 CIPK1（丝氨酸/苏氨酸蛋白激酶）同源，参与信号转导；参与代谢调节的 R011、R029 和 R032 分别与 GAPDH（甘油 3-磷酸脱氢酶），F-box 蛋白和 OMT（甲基转移酶）同源。总之，早期持续响应的基因主要参与转录和代谢调节。

3.4 讨论

本试验研究表明外源 ABA 能缓解辣椒的冷害症状（Guo et al.，2012）：未喷施 ABA 的植株出现严重的冷害症状，而 ABA 处理的植株叶片仅出现轻微伤害。因此，ABA 调节的基因表达在抗逆中起着重要作用。低温信号转导和 ABA 信号转导存在很多交叉。而且，遭受低温胁迫的辣椒由 ABA 调节的基因不同于以前报道的其他作物上 ABA 单一处理分离的基因（Matsui et al.，2008；Seki et al.，2002）。本试验试图分离 ABA 调节的辣椒耐寒相关基因。采用正反交文库构建，分离了 73 个 unigenes，为进一步分析抗逆性的生理机理和抗性植物的遗传转化提供资源。

3.4.1 分析电导率、叶绿素含量、净光合速率和气孔导度

在完整的植株中失水是低温胁迫最常见的症状，最终导致萎蔫。低温期间和之后出现的萎蔫归因于细胞膜破坏或正常流动状态过渡到不流动，凝固的状态（Wright，1974）。低温下电导率的增加是细胞膜破坏的一个指标（Bertin et al.，1996）。喷施 ABA 降低其增加，与 ABA 对冷害症状的保护作用变化一致。ABA 预处理缓解低温下总叶绿素含量的下降，与以前的研究结果一致（Kumar et al.，2008）。总叶绿素含量的下降可能由低温下产生的活性氧引起（Krause，1994），我们的研究结果也证实了这一点（低温显著增加辣椒 MDA 和 H_2O_2 含量）（Guo et al.，2012）。外源 ABA 也有效地缓解了低温下辣椒净光合速率的快速下降，这与 Jiang et al.（2012）在番茄上的报道结果一致。在辣椒遭受低温 3 h 时 ABA 使气孔导度，这是很多植物响应 ABA 反应的典型表现

(Mittelheuser and van Steveninck，1969)。

3.4.2 未知功能的基因

在 ABA 预处理和未处理的低温辣椒幼苗中总共分离具有不同表达水平的 73 个 unigenes，表明很多基因参与 ABA 调节的耐寒反应。Contigs 冗余率 76.6% 表明通过正反交文库大量的克隆和测序找到其他基因的可能性，特别是编码低丰度蛋白的基因。而且，73 个 unigenes 编码的蛋白参与代谢、防御反应、转录、信号转导等其他反应。正交库分离的大多数基因与转录有关（12.50%），表明在遭受低温胁迫的植株中，ABA 通过很多机制调节转录因子。

具有未知功能和低同源性的基因也占较大部分。正反交文库中未知功能基因占 30%。这些未知功能或新基因或许是抗逆候选基因（Baisakh et al.，2008）。对 4 个未知功能基因进行实时定量分析。在 ABA 预处理的低温辣椒幼苗 F012 和 F029 表达分别在低温 1 h 和 24 h 比对照高 3 倍，表明 ABA 对基因的诱导分别发生在早期和后期低温时期。相对于 0 h 低温，低温处理的幼苗 R002 和 R004 基因也显著上调，表明这些基因参与低温的适应能力。虽然这些基因在 ABA 调节低温反应中的功能仍不清楚，他们作为新的耐寒相关基因应该分离。

3.4.3 低温下 ABA 上调的运输相关基因

Ca^{2+} 泵参与耐盐性调节，内质网 Ca^{2+} — ATPases 表达量增加，比如番茄（Wimmers et al.，1992）和烟草（Perez-Prat et al.，1992）。在低温 12 h 后，对照 calcium-ATPase（F030）基因表达缓慢增加，表明 calcium-ATPase 参与后期低温反应。ABA 介导低温对气孔孔径的影响（Lee et al.，1993）。ABA 与保卫细胞质膜外膜上的识别位点结合，诱导信号转导，引起细胞质内钙离子增加，最终降低保卫细胞渗透势引起的气孔关闭（Assmann and Shimazaki，1999）。在 ABA 预处理植株上低温 3 h 后 calcium-ATPase 基因表达量迅速增加（对照 5 倍）。考虑到在 ABA 预处理的低温植株上气孔导度快速下降，这个结果表明 calcium-ATPase 可能参与低温下 ABA 介导的气孔关闭。也表明质膜 calcium-

ATPase 的表达调节发生在转录水平,与以前报道的保卫细胞一致 (Schiøtt and Palmgren, 2005),而且其调节也可发生在转录后水平 (Dietz et al., 2001)。

3.4.4 低温下 ABA 上调的转录相关基因

目前研究表明,一些 ABA 诱导植物特异的转录因子家族成员中含有 DNA 结合的 NAC 区域(NAM、ATAF 和 CUC)的,也参与 ABA 介导的信号转导反应。拟南芥 *ANAC019*、*ANAC055*、*RD26/ANAC072* 和 *ATAF1* 受 ABA 调节上调(Fujita et al., 2004;Tran et al., 2004)。辣椒 NAC 转录因子(F007)受 ABA 和低温影响上调。ABA 介导的干旱信号在转录和蛋白水平上诱导 NTL4(*ANAC053*)表达,NTL4 促进与耐旱性密切相关的活性氧产生(Lee et al., 2012)。依靠 ABA 的脱水信号转导上调 *RD26/ANAC072* 表达,其参与活性氧的解毒和防御反应(Fujita et al., 2004)。并且,拟南芥(Fujita et al., 2004;He et al., 2005)和水稻(Hu et al., 2008) *ANACs* 基因(响应 ABA 的 NACs)的过量表达增加对 ABA 的敏感性和提高植株的抗逆性。值得注意的是,遭受低温胁迫的辣椒植株 NAC 转录因子被 ABA 处理强烈诱导;与对照相比,在低温 1 h 和 24 h 其基因表达量增加 4 倍。考虑到外源 ABA 降低电导率,和 MDA 和 H_2O_2 含量 (Guo et al., 2012),进而提高辣椒的耐寒性,我们推测 ABA 调节的 NAC 转录因子启动某些基因协同提高辣椒的耐寒性,然而这些机制有待进一步研究。

MADS-box 基因编码重要的转录因子,参与植物生长和信号转导调节。温度敏感性是很多 *MADS-box* 基因的共同特征(Hemming and Trevaskis, 2011)。拟南芥至少有 14 个 *MADS-box* 基因受低温上下调节(Hannah et al., 2005)。ABA 预处理的低温植株 *MADS-box* 基因表达(F027)迅速达到峰值,而且 ABA 诱导其表达依靠低温,表明该基因转录水平的变化受低温影响。

3.4.5 低温期间 ABA 上调的代谢和防御相关基因

很多 ESTs 参与蛋白合成和能量代谢,这对植物上 ABA 介导的抗逆反应很有必要。大量证据表明翻译因子除了常规地负责蛋白合成外，在细胞中具有其他的功能（Hossain et al.，2012）。比如，拟南芥翻译延伸因子（EF2）突变体特异性地抑制其下游基因，进而降低植株的耐寒性（Guo et al.，2002）。低温诱导辣椒 *EF2* 基因表达，在大麦上也有类似报道（Faccioli et al.，2001）。在破坏种子休眠期间 EF2 蛋白表达受 ABA 下调，EF2 参与愈伤组织蛋白合成和细胞分裂（Pawłowski，2009）。然而，在遭受低温的辣椒上 ABA 在转录水平启动 *EF2* 表达，表明其参与 ABA 调节的耐寒性。

我们从正交库中分离编码光合作用相关的基因，如生成碳酸脱水酶（F026）和核酮糖羧化酶（F039）的基因。碳酸脱水酶通过影响电子传递为核酮糖羧化酶提供二氧化碳，在光合作用中起着重要的作用（Ignatova et al.，2011）。水稻碳酸脱水酶（OsCA1）转基因拟南芥幼苗耐盐性增强（Yu et al.，2007）。在遭受低温胁迫的辣椒上，ABA 迅速诱导碳酸脱水酶基因表达，而其在耐寒性方面所起的作用有待研究。低温胁迫下光合系统 II 的下降导致活性氧积累（Saibo et al.，2009）。而且，ABA 保护光合系统 II，进而降低低温引起的光抑制（Zhou et al.，2006），或许与编码核酮糖羧化酶小亚基的基因有关，也和叶绿素损失的缓解有关。

低温胁迫下辣椒的几个防御相关基因，如超氧化物歧化酶（F031）和几丁质结合蛋白（F008），受 ABA 上调表达。SOD 高效地歧化超氧阴离子生成过氧化氢。外源 ABA 诱导 *SOD* 表达，进而提高辣椒耐寒性，与以前研究的结果一致（Guo et al.，2012）。低温胁迫下辣椒编码几丁质结合蛋白的基因也受 ABA 上调表达。这个基因参与生物防御反应（Pushpanathan et al.，2012），表明生物和非生物胁迫引起的信号转导存在交叉过程。

3.4.6 低温期间 ABA 下调的基因

为了全面理解响应非生物胁迫的分子反应，分析响应低温反应过程中 ABA 下调和 ABA 上调基因一样重要。在反交库中，低温期间辣椒的几个与调节相关的基因受 ABA 下调表达，如编码 MBF

转录辅激活因子（R028）、F-box 蛋白（R029）和 CIPK（R030）。F-box 家族蛋白是泛素/蛋白水解途径中的关键成分，调节蛋白的选择性降解，从而在抗逆中起着重要作用（Paquis et al.，2011）。F-box 蛋白也是参与激素信号转导的关键因子（Tan and Zheng，2009）。在辣椒幼苗上，单一的 ABA 和低温处理上调 *F-box* 基因表达，与其他的研究结果一致（Paquis et al.，2011；Vogel et al.，2005）。拟南芥 F-box 蛋白突变体（*DOR*）耐旱性增强（Zhang et al.，2008）。并且，转基因水稻 *F-box* 基因的过量表达降低植株的抗逆性（Yan et al.，2011）。ABA 介导 *F-box* 基因下调或许有助于辣椒的耐寒性提高。拟南芥 CIPK1 是钙信号转导途径的主要成分，参与盐和渗透胁迫转导，进而调节依靠 ABA 和不依靠 ABA 的逆境反应（D'Angelo et al.，2006）。MBF 是一个转录辅激活子，通过连接转录因子和顺式作用元件来调节转录激活。拟南芥 *MBF1* 基因与抗逆性有关，在依赖 ABA 抑制种子发芽过程中起着负调节作用（Mauro et al.，2012）。低温诱导 *MBF* 和 *CIPK* 基因表达；与对照相比，ABA 预处理的低温植株其表达量下降，分别是对照的 0.105 倍（3 h）和 0.31 倍（1 h）。这些结果表明在遭受低温胁迫的辣椒上，*MBF* 和 *CIPK* 基因参与快速且复杂的基因表达调节。

我们从反交库中获得脱水素基因（R031），参与防御反应。拟南芥脱水素基因表达在响应低温和 ABA 反应中变化不同（Nylander et al.，2001）。遭受低温胁迫的辣椒脱水素基因表达增加，而非 ABA 处理。代谢相关基因（R032）编码木质素合成蛋白，也从反交库中获得。在逆境胁迫中植物木质素的合成发生变化（Moura et al.，2010）。比如，在干旱胁迫下，参与编码木质素合成的 *OMT* 基因表达增加（Yang et al.，2006），与耐旱性有关（Yoshimura et al.，2008）。单一的 ABA 和低温处理诱导 *OMT* 基因表达可能参与木质素合成。而且，我们又分离出另一个代谢相关的基因 *GAPDH*，编码糖酵解途径中的关键酶。目前研究表明 GAPDH 是各种细胞中的多功能蛋白。比如，拟南芥 *AtGAPC-1*

直接与 H_2O_2 互作介导活性氧信号转导（Hancock et al.，2005）。转基因水稻 *OsGAPC3* 过量表达通过调节 H_2O_2 含量来提高作物的耐盐性（Zhang et al.，2011）。植物 *GAPDH* 基因在各种逆境胁迫下诱导表达，比如蘑菇低温胁迫（Jeong et al.，2000）和水稻 ABA 处理（Zhang et al.，2011）。在单一的 ABA 和低温处理下辣椒 *GAPDH* 基因（R011）表达迅速增加。然而 ABA 预处理的低温植株 *GAPDH* 基因表达基本没有发生变化。考虑到 ABA 有效地降低低温下辣椒冷害指数、MDA 和 H_2O_2 含量，这些结果表明外源 ABA 不足以启动 *GAPDH* 基因表达来预防低温下活性氧的过量积累。

与对照相比，ABA 预处理的低温辣椒存在很多胁迫相关的基因。然而，有些逆境标记基因如 *RD29A/B*、*KIN1*、*KIN2*、*CBF1*、*RAB* 没有从正反交文库中分离得到。这可能与以下原因有关：克隆点测序的有限性，研究材料种类的差异性，胁迫处理的强度和时间，取样时间和方法的不同以及正反交文库的交叉性。而且 ABA 和低温信号转导过程存在交叉点，可能某些逆境标记基因同时存在于正反交文库。

3.5 结论

本试验通过分析 ABA 介导的基因表达变化，最终分离出 ABA 调节的低温抗性相关基因。在遭受低温胁迫的辣椒幼苗上总共分离出 73 个 ABA 调节的 unigenes。其中 37 个 unigenes 根据其同源基因在防御反应中的作用分成不同的功能。在低温下受 ABA 调节的这些基因或许与 ABA 在植物的生长发育和逆境适应中起着重要的作用有关。本试验有助于深入分析在 ABA 预处理和清水预处理植物上几种耐寒机制的相关性。ABA 调节的低温抗性相关基因编码蛋白，其功能和靶基因有待进一步研究。而且这些基因的表达和翻译调节机制仍不清楚。作为转基因植物的候选基因，NAC 和 MBF 转录因子通过对耐寒性相关的靶基因进行时空性调节，参与 ABA 介导的基因表达需要进一步研究。

4 辣椒 CaNAC2 转录因子的功能分析

低温、干旱和盐胁迫严重影响植株的生长和发育。当植物受到胁迫时，诱导很多基因表达，引起一些生理生化变化来躲避逆境造成的危害（Thomashow，1999；Shinozaki et al.，2003）。一般认为，转录因子的启动通常会诱导很多逆境相关功能基因的表达。关于过量表达参与逆境反应的转录因子能够提高植株的抗逆性的报道有很多，如 *DREB/CBF*（Liu et al.，1998），*ABF2*（Kim et al.，2004），*OSISAP1*（Mukhopadhyay et al.，2004），*SNAC1*（Hu et al.，2006）等。

NAC 转录因子是植物特有的，同时也是一个很大的基因家族。此家族蛋白含有高度保守的 N 端 DNA 结合结构域和变化多样的 C 端转录激活域（Ernst et al.，2004）。目前水稻和拟南芥已分别发现 151 和 117 个 NAC 成员（Nuruzzaman et al.，2010）。越来越多的研究表明：NAC 蛋白不仅调节植物的生长发育，如分生组织发育、花叶器官发育、侧根发育、细胞周期调控和次生细胞壁的合成等（Guo and Gan，2006；Hibara et al.，2003；Nobutaka et al.，2007；Kim et al.，2006；Souer et al.，1996；Xie et al.，2000；Zhong et al.，2006），而且参与激素调节、生物和非生物胁迫反应如 ABA、乙烯、干旱、高盐、低温等（He et al.，2005；Hu et al.，2008；Tran et al.，2004）。比如 *AtNAC1* 参与生长素信号转导促进侧根生长（Xie et al.，2000）；*SND1* 调节拟南芥茎的次生壁合成（Zhong et al.，2006）；在拟南芥上，3 个 *NAC* 基因 *ANAC019*、*ANAC055* 和 *ANAC072* 受 ABA、干旱、盐胁迫或低温诱导表达，并且过量表达这些基因的植株抗逆性增强（Fujita et al. 2004；Tran et al.，2004）；同样水稻 *SNAC1* 和 *SNAC2* 也受 ABA、低温、盐胁迫和干旱诱导，而且该基因的过量表达株系表现较高的抗逆性，如失水、高盐或低温胁迫（Hu et al.，2006，2008）。高盐和 ABA 诱导 *AtNAC2* 基因表达，过量表达转基因株系改变侧根的生长发育和提高植株的耐盐性（He et al., 2005）。以上结果表明 NAC 基因家族的功能多样性。目前报道，在不同的生物和非生物胁迫下，茄科作物 NAC 基因表达变化不同：马铃薯 *StNAC* 受疫病菌诱导（Collinge et al., 2001），*SlNAC1*、*SlNAM1* 受高盐诱导（Yang et al., 2011），*SlNAC3* 受干旱、高盐和 ABA 抑制表达（Han et al., 2012）。辣椒 *CaNAC1* 受病原菌和 SA 显著诱导（Oh et al., 2005）。

为了研究 ABA 调节的耐寒机制，我们构建了抑制消减杂交文库（SSH），获得与 NAC 蛋白高度同源的 EST（Guo et al., 2013），以其为探针，从辣椒叶片中克隆基因全长，命名为 *CaNAC2*，并对其在非生物胁迫中所起的生物学功能进行研究。具体内容为①*CaNAC2* 转录因子的定位和转录活性分析；②该基因在辣椒不同组织和非生物逆境胁迫下的表达模式分析，如 SA、高盐、甘露醇和汞胁迫；③发生 *CaNAC2* 沉默的植株遭受非生物胁迫的抗性分析。

4.1 材料与方法

4.1.1 主要材料及处理方法

本研究拟采用的辣椒材料为 P70 栽培种，由西北农林科技大学辣椒课题组提供。辣椒生长条件参照第 3 章材料与方法描述。植株生长 6~7 片真叶，进行 ABA、SA、低温、盐和干旱处理，每个处理重复三次。0.57 mmol/L ABA 和低温处理参照第 3 章材料与方法描述。SA 处理，5 mmol/L SA 进行叶面喷施长势一致的辣椒材料；盐胁迫，首先将辣椒从基质中拔出（勿伤到根系），放到水中适应 1 h，然后转移到 300 mmol/L NaCl 溶液中浸泡；渗透胁迫，同样将辣椒从基质中拔出，冲洗干净，放到水中适应 1 h，然后浸泡在 300 mmol/L 甘露醇溶液中，所有处理在 0、1、3、6、12、24 和 48 h 进行取样，液氮速冻后于 −70℃ 条件保存备用。

4.1.2 NAC 基因分离和序列分析

利用 SSH 正交库 EST（登录号 JZ198750）作为探针，在

Blastn (http://blast.ncbi.nlm.nih.gov/Blast.cgi) EST 数据库进行序列同源性比较，搜索物种为茄科（Solanacae）；根据比对的高同源 EST 序列利用 ContigExpression 9.1 软件拼接全长，同 Primer 5.0 软件设计全长引物进行 PCR 扩增，即同源克隆法（Zhu et al., 2011）。*CaNAC2* 全长引物为：F: 5′-TTCTTCCTTTTTATTTATCTGTG-3′；R: 5′-GGTACAGGTTATAGCGAAACTACG-3′。

根据基因推导的氨基酸序列进行蛋白同源性（NCBI blastx）、蛋白结构（DNAMAN 6.0）及亚细胞定位分析（ProComp v 8.0, http://linux1.softberry.com/berry）。利用 MEGA 5 软件进行分子进化分析，采用邻近法（Neighbor-Joining，N-J）构建系统发育树，1 000 次重复，节枝点上的数值代表可信度。

4.1.3 表达载体的构建

（1）病毒沉默 *CaNAC2* 载体的构建

①重组 pTRV2-*PDS* 载体的构建。病毒载体包括 pTRV1 和 pTRV2 两个独立的质粒载体。目的基因插入 pTRV2 多克隆位点，pTRV1 主要负责病毒在植物组织中的转运。根据已发表的辣椒 *PDS* 基因序列（GenBank NO. X68058），按照王军娥等（2013）的方法进行构建 pTRV2-*PDS*。

②重组 pTRV2-*CaNAC2* 载体的构建。为了获得更好的基因沉默效果，我们设计引物时需要满足以下条件：插入的目的片段大小为 150~500 bp（Thomas et al., 2001）。为此，我们在 *CaNAC2* 特异区域设计引物以保证基因沉默的特异性，片段长度为 447 bp。将相应的特异片段插入到病毒载体 pTRV2，构建流程如图 3-19 所示：

所用引物：

VCaNAC2-F:5′-CCGACCTCTGACGTTTGTTTG-3′

VCaNAC2-R:5′-CGCGGATCCAGTTTCCTCAAGTCCTCGTTC-3′（下游引物添加 *Bam* HI 酶切位点和保护碱基，下划线为 *Bam* HI 酶切位点）。

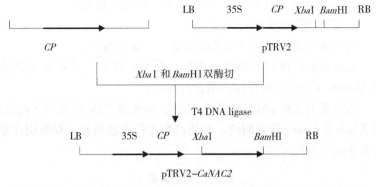

图3-19 pTRV2-*CaNAC2* 构建流程

a. PCR扩增 *CaNAC2* 特异片段。

b. PCR扩增产物在1‰琼脂糖凝胶上进行电泳检测，紫外灯下将目的条带进行切胶回收，按照胶回收试剂盒（Tiangen，China）的操作说明回收PCR产物。

③将回收纯化的目的基因与pMD19-T载体连接，按照试剂盒（Takara，Japan）操作说明进行连接。连接体系见表3-7，混匀后，在16℃条件下连接2 h以上。

表3-7 pMD19-T载体连接体系

反应组分	加样量（μL）
溶液 I	2.5
pMD19-T载体	0.5
目的片段	2.0
总体积	5.0

④连接产物采用冻融法转化感受态大肠杆菌DH5α 具体步骤：

a. 连接产物加入50 μL DH5α感受态，冰上放置30 min。

b. 42℃水浴热激90 s，立即冰上冷却2 min。

c. 加200 μL LB液体培养基，然后37℃，180 r/min振荡培养1 h。

d. 菌液涂布到含有 50 mg/L Amp LB 的平板上。

e. 37℃倒置培养 14~16 h。

⑤大肠杆菌 DH5α 感受态的制备——氯化钙法。

⑥碱裂解法提取大肠杆菌 DH5α 质粒，按照质粒提取试剂盒（Tiangen，China）的操作说明进行提取。

⑦将重组质粒 pMD19-T-*CaNAC2* 和病毒载体 pTRV2 分别进行 *Xba*I 和 *Bam*HI 双酶切，酶切产物进行凝胶回收。双酶切体系见表 3-8。

表 3-8 酶切体系

反应组分	加样量（μL）
10×Tango 缓冲液	3.0
*Xba*I	1.0
*Bam*HI	1.0
pMD19-T-*CaNAC2*（或病毒载体 pTRV2）	10.0
无菌水	15.0
总体积	30.0

混匀后，37℃酶切反应 4 h 左右。

⑧将目的基因与酶切后的载体进行凝胶回收纯化，然后使用 T4 DNA ligase（Clontech，Japan）连接。连接体系见表 3-9。混匀后，16℃连接 2 h 以上。

表 3-9 pTRV2 载体连接体系

反应组分	加样量（μL）
5×T4 DNA 连接酶缓冲液	1.0
T4 DNA 连接酶	0.5
病毒载体 pTRV2	0.5
目的基因	3.0
总体积	5.0

第3章 辣椒对低温胁迫的响应与其低温抗性相关基因的克隆和功能分析

⑨病毒载体 pTRV2-*CaNAC2* 转化感受态农杆菌 GV3101。

具体步骤:

a. 取 100 μL 感受态细胞,加入质粒 0.8 μL,混匀后冰上放置 30 min。

b. 将离心管迅速放入液氮中冻 90 s。

c. 立即将离心管加入预加温至 37℃ 的水浴中,热休克 5 min。

d. 快速将离心管插入冰上,使细胞冷却 2 min。

e. 加入 1 mL LB 培养基,28℃,180 r/min 摇至 4~6 h。

f. 4℃,4 000 r/min 离心 5 min,收集细胞,用 50 μL LB 培养基重悬细胞沉淀,然后涂布在含有抗生素的培养基上(10 mg/L Rif+50 mg/L Gen+50 mg/L K$^+$)。

g. 平板置于室温至菌液被吸收,28℃ 倒置培养。

⑩农杆菌 GV3101 感受态制备。

a. 挑取 GV3101 单菌落于 LB 液体培养基(10 mg/L Rif+50 mg/L Gen+50 mg/L K$^+$),28℃,200 r/min 培养 24~32 h。

b. 取 2 mL 菌液接种于 100 mL 新鲜含有相应抗生素的 LB 液体培养基中(1∶50),继续 28℃ 振荡培养至 OD$_{600}$ 0.5~0.8(10~12 h)。

c. 置于预冷的 50 mL 离心管中,冰上放置 30 min,4℃,5 000 r/min 离心 10 min,弃上清。

d. 用预冷的 5 mL 20 mmol/L CaCl$_2$ 重悬细胞。

e. 4℃,5 000 r/min 离心 5 min,弃上清。

f. 倒置,加 1 mL 预冷的 20 mmol/L CaCl$_2$ 重悬细胞,再加预冷的无菌 50% 甘油约 425 μL,至终浓度为 15%,即成感受态细胞。

g. 将感受态细胞 100 μL 分装到 1.5 mL 离心管中,液氮速冻,立即置于 −80℃ 保存或冰上进行转化。

(2)亚细胞定位载体的构建 将去除终止密码子的 *CaNAC2* 开放阅读框插入到表达载体 pBI221-GFP 上。流程如图 3-20 所示。

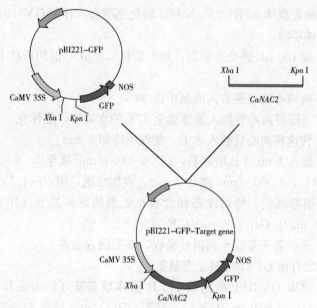

图 3-20 亚细胞定位载体的构建

所用引物：

GFP-CaNAC2-F：5′-GGAAATGGAGCAAGAAGGA-3′；

GFP-CaNAC2-R：5′-CGGGGTACCACCATTTCGGACGAGTTTC-3′（下划线为 KpnI 酶切位点）。

用 XbaI 和 KpnI 限制性内切酶酶切 pMD19-T-CaNAC2 载体和空表达载体 pBI221-GFP。酶切体系见表 3-10。

表 3-10 酶切体系

反应组分	加样量（μL）
10×Bam HI 缓冲液	3.0
XbaI	1.0
KpnI	1.0
pMD19-T-CaNAC2（或空载体 pBI221-GFP）	10.0
无菌水	15.0
总体积	30.0

第3章 辣椒对低温胁迫的响应与其低温抗性相关基因的克隆和功能分析

酶切产物进行凝胶回收纯化,连接目的片段 CaNAC2 和 pBI221-GFP 载体均参照以上提到的方法。

(3) 分析转录激活活性的酵母表达载体构建 转录活性分析的酵母表达载体 pGBKT7,该载体中含有酵母 GAL4 DNA 结合域。依据与其他同源基因的氨基酸比对结果,在 CaNAC2 开放阅读框外设计上游引物和下游引物,构建载体 pGBKT7-CaNAC2 流程如图 3-21 所示:

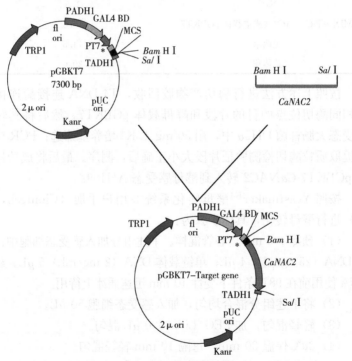

图 3-21 转录激活载体的构建示意图

所用引物:

TCaNAC2-F:5′-TTCTTCCTTTTTATTTATCTGTG-3′

TCaNAC2-R:5′-GGTACAGGTTTATAGCGAAACTACG-3′

将 CaNAC2 全长克隆到 pMD19-T 载体上。按照以上方法进行 PCR 扩增,PCR 产物连接到 pMD19-T 载体,然后用 Bam HI

和 Sal I 限制性内切酶进行 37℃双酶切 3 h。同时，空载体 pGBKT7 也进行相同的双酶切。酶切体系见表 3-11。

表 3-11 酶切体系

反应组分	加样量（μL）
10×Bam HI 缓冲液	3.0
BamHI	1.0
SalI	1.5
pMD19-T-CaNAC2（或空载体 pGBKT7）	10.0
无菌水	14.5
总体积	30.0

按照上述方法进行酶切产物胶回收，T4-DNA 连接酶连接带有相同酶切位点的目的片段和酵母载体 pGBKT7，然后将其转入感受态大肠杆菌 DH5α 中，用 50 mg/L K$^+$ 培养基筛选；PCR 检测和提取质粒酶切检测验证片段大小正确后，测序。最后将成功构建的 pGBKT7-CaNAC2 转入到酵母感受态 AH109。

按照 Yeastmaker^{TM} 酵母转化系统 2 用户手册（Clontech，日本）进行酵母转化，具体步骤为：

（1）预冷 1.5 mL 离心管混样。下述组分加入感受态细胞中：质粒 DNA（25 ng/μL）4 μL；单链载体 DNA（2 mg/mL）5 μL。载体 DNA 使用前在 98℃条件下变性 10 min 迅速插冰上待用。

（2）将上述组分混合均匀，加入感受态细胞 50 μL。

（3）轻轻混匀，加 PEG/LiAc 500 μL 混匀。

（4）30℃保温 30 min，每隔 10 min 轻轻混匀。

（5）加 DMSO 20 μL，混匀。

（6）42℃热激 25 min，每隔 5 min 混匀。

（7）室温放置 2 min，瞬时离心 15 s，收集菌体，用 1 mL YPD plus medium 培养基重悬菌体。

（8）30℃，180 r/min 摇 1 h，短暂离心收集菌体。最后用 1 mL 0.9%（w/v）氯化钠溶液重悬菌体，按 1：10 稀释后，涂布于固体筛选培养基 SD/-Trp，SD-Leu/-Trp 和 SD-Ade/-His/-Leu/-Trp 上，30℃培养 3 d。

酵母感受态 AH109 制备的方法参考朱自果的方法（2012）。

4.1.4 亚细胞定位分析

采用 BiolisticPDS-1000/He（BioRad）基因枪轰击洋葱表皮进行亚细胞定位分析参考朱自果的方法（2012）。

4.1.5 辣椒 CaNAC2 转录激活活性分析

为了分析 CaNAC2 是否具有转录活性，将连接 CaNAC2 全长的重组质粒转化至酵母 AH109 上。然后，在缺陷型培养基 SD/-Trp、SD-Leu/-Trp 和 SD-Ade/-His/-Leu/－Trp 上，30℃培养 3～5 d。

4.1.6 辣椒 CaNAC2 在非生物胁迫下的表达模式分析

植株生长到 6～7 片真叶，进行 ABA、SA 及盐渗透处理。RNA 提取，单链 cDNA 合成，实时定量均参考第 3 章。

4.1.7 VIGS 技术分析辣椒 CaNAC2 沉默

辣椒 CaNAC2 沉默参考 Choi et al.（2007）、Liu et al.（2002）、Wang et al.（2013）的叶片注射法，略加修改，注射苗龄是子叶完全展开时，用针头在靠近主脉处扎小孔。具体操作：

a. 取携带目的片段质粒的农杆菌（pTRV1、pTRV2-CaNAC2/PDS）在 LB 培养基（10 mg/L Rif＋50 mg/L Gen＋50 mg/L K$^+$）上划板，28℃培养 2 d。

b. 挑取单菌落，用含有相应抗生素的 LB 液体培养基摇菌，200 r/min，28℃摇 16～18 h。

c. 取菌液按 1∶25 加入诱导培养基（含有 10 mg/L Rif＋50 mg/L Gen＋50 mg/L K$^+$＋200 μmol/L 乙酰丁香酮），28℃摇 16～24 h，使菌液 OD$_{600}$ 值为 0.5～0.8。

d. 收集菌体，5 000 r/min 离心 10 min，用相同体积的 MES 悬浮液重悬。再次离心，收集菌体，用 1/2 体积的 MES 悬浮液重悬。

e. 在含有 pTRV1 质粒的悬浮液里加入乙酰丁香酮（终浓度 400 μmol/L），室温放置 3～5 h。

f. 将 pTRV1 分别和 pTRV2-*CaNAC2*、pTRV2-*PDS* 等量混合，此时乙酰丁香酮终浓度为 200 μmol/L，用于注射。

接种后植物放置在光照培养箱中（22℃、60%空气湿度）暗培养 2 d，之后，转入 22℃、16 h/8 h 光周期下正常生长。接种前 1 d 浇水，接种后 3 d 控水以增加发病率。观察注射植株的表型变化，如沉默 *PDS* 的辣椒幼苗新长出叶片是否变白。注射后 3～4 周，对发生基因沉默的植株进行非生物胁迫处理。

（1）辣椒 *CaNAC2* 沉默效率的分析 根据沉默 *PDS* 的辣椒幼苗新长出叶片变白进行初步判断辣椒品种 P70 可能出现的基因沉默时期，取叶位相同的叶片进行 RNA 提取，然后通过实时定量分析 *CaNAC2* 的沉默效率。

（2）ABA 和低温对 *CaNAC2* 沉默的植株 MDA 和 H_2O_2 含量的影响 对发生 *CaNAC2* 沉默的植株分别进行叶面喷施 0.57mmol/L ABA 和清水，72 h 后放置于低温 6℃条件，光强为 6 000 lx，处理 24 h 取样进行生理指标测定。MDA 和 H_2O_2 含量的测定方法参考第 3 章第 2 节。

（3）盐和渗透胁迫对 *CaNAC2* 沉默的植株表型影响 取发生 *CaNAC2* 沉默植株的叶片（1 cm^2），分别放在 300 mmol/L NaCl 和 300 mmol/L 甘露醇溶液中，培养条件为：25℃、6 000 lx 光强，16 h/8 h 光周期放置 3 d，观察离体叶片的表型变化。

4.2 结果与分析

4.2.1 辣椒 *CaNAC2* 全长 cDNA 的分离

以 SSH 正交库中筛选的 NAC 同源 EST（F007）为探针，从 GenBank 中比对选取 50 个茄科高同源 ESTs 序列，通过 Contig-Express 软件进行重叠拼接分析，获得推导的核酸序列包含完整的开放阅读框。然后，在拼接序列两端设计引物进行 PCR 扩增，以 P70 栽培种 cDNA 为模板，克隆基因全长，命名为 *CaNAC2*。其全长序列为 1490 bp（GenBank 登录号：JX402928），开放阅读框为 1230 bp，编码 409 个氨基酸，分子量为 45.8 KD，等电点为 5.34。

第3章 辣椒对低温胁迫的响应与其低温抗性相关基因的克隆和功能分析

基因测序获得的 cDNA 全长及其氨基酸序列如图 3-22。

```
1 TTCTTCCTTTTTATTTATCTGTGTATGTATATATTGTTTTTGGTGTGGGTTTC
54 TTGAAATAGAACCCCTGAATTCTCATTCTCTTGGGTTT TCAGGCTTGGAA
104 ATGGAGCAAGAAGGAGCACTTGTTTTAGTACCGGCTGCAGTGGTCGCACGCCGAATAGGGTAGTGCCGCCACCACCACCACGACG
     M  E  Q  E  G  A  L  V  L  V  P  A  A  V  V  A  T  P  N  R  V  V  P  P  P  P  P  T
191 TCGTTAGCGCCAGGTTTTCGGTTTCATCCAACTGATGAGGAGTTGGTGAGGTACTATAAGGAGGAAGGCTTGTGCAAAGCCATTT
     S  L  A  P  G  F  R  F  H  P  T  D  E  E  L  V  R  Y  Y  L  R  R  K  A  C  A  K  P  F
278 CGATTTCAGGCTGTTGCTGAAATTGATGTTTACAAATCTGAACCTTGGGAACTTGCTGAGTATTCATCTCTGAAGACCAGAGATTTA
     R  F  Q  A  V  A  E  I  D  V  Y  K  S  E  P  W  E  L  A  E  Y  S  S  L  K  T  R  D  L
365 GAGTGGTACTTCTTCTCCCCAGTTGATAGGAAGTATGGAAATGGATCTCGACTGAATCGAGCCACAGGGAAGGGTTACTGGAAAGCA
     E  W  Y  F  F  S  P  V  D  R  K  Y  G  N  G  S  R  L  N  R  A  T  G  K  G  Y  W  K  A
452 ACTGGGAAGGATCGTCCTGTTCGCCACAAATCTCAGACCATTGGAATGAAGAAAACAGCTTGTCTTCCATAGTGGTCGAGCTCCTGAT
     T  G  K  D  R  P  V  R  H  K  S  Q  T  I  G  M  K  K  T  L  V  F  H  S  G  R  A  P  D
539 GGTAAGAGAACTAATTGGGTGATCCATGAGTACAGACTTGCCGATGAAGAATTGGACAGAGCTGGAGTTGTCGCAGGATGCTTTTGTA
     G  K  R  T  N  W  V  M  H  E  Y  R  L  A  D  E  E  L  D  R  A  G  V  V  A  G  D  A  F  V
626 CTTTGTAGAACTCTTCCAAAAGAGTGGATTAGGACCTCCTAACGGTGACAGATATGCGCCATTTATTGAGGAGGAATGGGATGACGAC
     L  C  R  I  F  Q  K  S  G  L  G  P  P  N  G  D  R  Y  A  P  F  I  E  E  E  W  D  D  D
713 ACAGCTCTCATGGTTCCGGAGGAGAGGCAGAGGATGATGTGGGCAATGGTGATGAAGCACAAGTCGAGGGACATGACCTTGACCAG
     T  A  L  M  V  P  G  G  E  A  E  D  D  V  G  N  G  D  E  A  Q  V  E  G  H  D  L  D  Q
800 CTGGCTGTTGGATCGAGCAAGCAGATGCTCTGGGAAAGGCCCCTTGCCAAAGTGAAGACTGGCTGAGCCCTCTCTGGTGACTTTCA
     L  A  V  G  S  S  K  Q  D  A  L  G  K  A  P  C  Q  S  E  N  L  A  E  P  R  P  L  T  F
887 GTTGCAAGAGGGAAAGGTCTGAAGAACTCGAACCCCTCTCCTTAGCTCAAAGCAAAAGATCAAGCATGATGATCTAGCTCTAGC
     V  C  K  R  E  R  S  E  E  L  E  P  L  S  L  A  Q  S  K  R  S  K  H  D  D  P  S  S  S
974 CATGCAAATGGTTCGGAAGATTCAATCACTACTTGACCAAAGATCCAACCAAGATTCTACTTCTCAGCAGGATCCCCCAACTCTTA
     H  A  N  G  S  E  D  S  T  T  S  Q  Q  D  P  P  T  L  M  T  T  S  Y  S  P  A  L  L
1061 ACATTCCCCGTGTTAGAACCCCTGAACCAAAAGAAAGCCAACCTTCTAACCCTCTTACTTTTGACTCTTCCAACCTTGAAAAATCT
     T  F  P  V  L  E  P  P  E  P  K  E  S  Q  P  S  N  P  L  T  F  D  S  S  N  L  E  K  S
1148 GTTCCTCCAGGTTACCTGAGGTTCATTAGCAATTTAGAGAATCAGATACTAATATGTGTCATCGAGAAGGAGACGGCAAGGATTGAA
     V  P  P  G  Y  L  R  F  I  S  N  L  E  N  Q  I  L  N  V  S  M  E  K  E  T  A  R  I  E
1235 GTGATGAGAGCTCAAGCTATGATCAACGTCCTTCTTCAATCGATCGATCTCTTGAACAAGGAGAACGAGGACTTGAGGAAACTCGTC
     V  M  R  A  Q  A  M  I  N  V  L  Q  S  R  I  D  L  L  N  K  E  N  E  D  L  R  K  L  V
1322 CGAAATGGTTAGCTGGTAACTATCTCCATGTAAGTCTAGTTTTTGTAGTTGTCGGTGGAGGCCCCCCCCCTTGT
     R  N  G  *
1394 ATCGGAGTCCCTCCATTCTTGAACTGTCTTTAGTAGGAGTAAGATACTGGC TATGTAATAAACGTTATT
1464 TACGTAGTTTCGCTATAAACCTGTACC
```

图 3-22 辣椒 *CaNAC2* cDNA 及其推导的氨基酸序列

4.2.2 辣椒 *CaNAC2* 推导的蛋白结构域及系统进化树分析

与其他物种的氨基酸同源比对结果表明：辣椒 CaNAC2 与矮牵牛 NH4 转录因子（AAM34767.1）、茄科夜来香 NAC（AFP93563.1）、番茄 NAC 类蛋白（XP_004239709.1）同源性高，同源率分别为 84%、76% 和 73%；而与拟南芥 ANAC053（NP_566376.1）、ANAC078（At5g04410）和 NAC2（AAF09254.1）同源性低，分别为 28%、27% 和 26%。为了分析 *CaNAC2* 编码的氨基酸结构特点，利用 DNAMAN 软件进行与其他物种的 NAC 蛋白序列比较，CaNAC2 N 端编码完整的 NAC 结构域（A、B、C、D、E），此外还含有一个核定位信号区域（双箭头处）（图 3-23）。其 C 端富含简单丝氨酸的重复序列（粗下划线），这是植物转录激活区域的共同特征之一。C 端含有大量重复的 α-螺旋结构（357-409 aa），不含有跨膜结构，这与其他的 NAC2 亚家族成员不同，如 ANAC053 和 ANAC078。

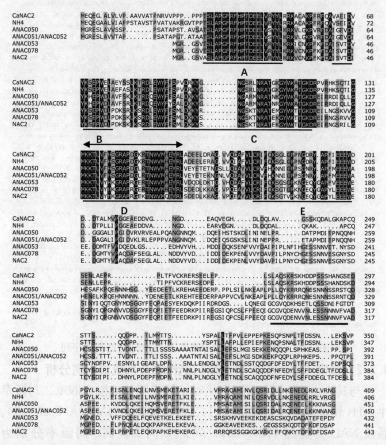

图 3-23　辣椒 CaNAC2 与其他 NAC 蛋白的氨基酸序列比对

横线处为 NAC 蛋白的 5 个保守结合位点（A-E），粗下划线处为 C 端丝氨酸的重复序列，双箭头处为核定位信号区域。其他物种的 NAC 蛋白为：矮牵牛 NH4（AAM34767.1），拟南芥 ANAC050（XP_002882675.1），ANAC051/52（XP_002882676.1），ANAC053（NP_566376.1），ANAC078（At5g04410），NAC2（AAF09254.1）

第3章 辣椒对低温胁迫的响应与其低温抗性相关基因的克隆和功能分析

利用 MEGA5 软件进行 CaNAC2 与其他已报道的 NAC 蛋白系统进化分析,结果如图 3-24 所示。根据 Ooka et al.（2003）的 NAC 家族分类方法,辣椒 CaNAC2 属于 NAC2 亚家族。与 SlNAC (HM053568)，ANAC053（NP_566376.1）属于同一亚家族。

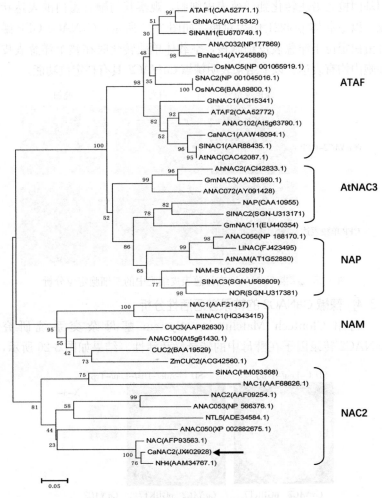

图 3-24 辣椒 CaNAC2 转录因子的进化树分析

4.2.3 辣椒 CaNAC2 的亚细胞定位分析

CaNAC2 蛋白分析表明其含有一个核定位信号区域,且使用 ProCompv 8.0 软件预测该蛋白定位于细胞核。为了研究 CaNAC2 在植物细胞中的定位,我们构建了 35S::CaNAC2-GFP 融合蛋白,采用基因枪轰击法转化到洋葱表皮细胞,观察其与融合蛋白的表达状况。以空载体 pBI221 作为对照。如图 3-25 所示,CaNAC2-GFP 融合蛋白定位于洋葱表皮细胞核,空载体单独转化时在整个洋葱表皮细胞内均有表达。以上结果表明辣椒 CaNAC2 具有核定位功能。

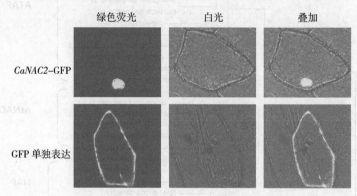

图 3-25 辣椒 CaNAC2 在洋葱表皮细胞中的亚细胞定位分析

4.2.4 辣椒 CaNAC2 的转录激活活性分析

利用 Clontech Matchmaker System 酵母杂交系统研究 CaNAC2 转录因子在酵母中的转录激活活性。结果如图 3-26 所示,

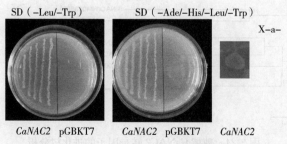

图 3-26 CaNAC2 在酵母中的转录激活分析
CaNAC2 代表融合蛋白(pGBKT7-CaNAC2);pGBKT7 为空载体对照

CaNAC2 全长序列与 GAL4 DNA 结合域融合在二缺 SD（-Leu/-Trp）培养基和四缺 SD（-Ade/-His/-Leu/-Trp）培养基上都能生长，并在涂有 X-α-gal 的培养基上变蓝，而对照（空载体）在相应的二缺和四缺培养基上不能生长，表明 CaNAC2 在酵母中具有转录激活功能。

4.2.5 辣椒 CaNAC2 的表达模式分析

（1）*CaNAC2* 在辣椒不同组织的表达分析　qRT-PCR 结果表明 *CaNAC2* 在辣椒不同组织中均有表达（图 3-27）。*CaNAC2* 表达量在根、幼嫩和种子中较高。

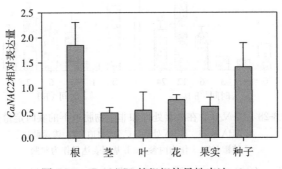

图 3-27　*CaNAC2* 的组织特异性表达

（2）*CaNAC2* 响应逆境胁迫反应的表达分析　为了分析在非生物胁迫和外源激素处理下辣椒 *CaNAC2* 的表达模式，本试验对辣椒幼苗进行盐胁迫、干旱、ABA 和 SA 处理，使用 qRT-PCR 检测 *CaNAC2* 在不同处理下的表达变化。ABA 处理 *CaNAC2* 表达量逐渐增加，48 h 上升到最大（高于对照 3 倍）（图 3-28A）；SA 处理显著抑制其表达量，1 h 和 3 h 基因表达量分别为 0 h 的 0.065 和 0.083 倍（图 3-28B）。高盐胁迫诱导 *CaNAC2* 表达（图 3-28C）：3 h 期间其表达量略微下降，6 h 之后增加，12 h 达到峰值。甘露醇 1 h 处理迅速抑制 *CaNAC2* 表达，3 h 后下降逐渐缓解，但其表达量整体低于对照水平（图 3-28D）。

4.2.6　VIGS 技术分析辣椒 CaNAC2 沉默的功能

（1）辣椒 *CaNAC2* 特异片段的分离　根据 DNAMAN 分析的

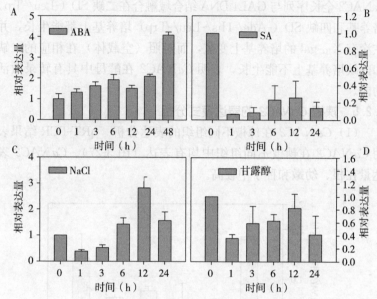

图 3-28 *CaNAC2* 在激素处理和非生物胁迫下的实时定量分析
A. ABA 处理　B. SA 处理　C. 盐胁迫　D. 渗透胁迫。
辣椒 *UBI-3* 作为内参，0 h 基因表达量作为对照

序列结果，在 *CaNAC2* 特异区域设计一对特异性引物。从辣椒 P70 叶片提取总 RNA，反转录成单链 cDNA，以其为模板进行 PCR 扩增，扩增产物长度为 447 bp。将其连入 pMD19-T 载体进行测序，测序结果与全长基因的特异区域完全一致。

（2）病毒沉默载体的构建　取测序正确的 pMD19-T-*CaNAC2* 质粒和空载体 pTRV2：00 分别进行 *Xba* I 和 *Bam* HI 双酶切。回收相应大小的片段，进行 16℃过夜连接，转化感受态 *E. coli* DH5α。酶切鉴定大肠杆菌转化子质粒，片段大小与预期长度一致，说明病毒载体 pTRV2-*CaNAC2* 已成功构建（图 3-29）。然后，将转化子质粒转入农杆菌感受态 GV3101。菌液 PCR 检测转入 pTRV2-*CaNAC2* 的菌株，以空载体 pTRV2：00 质粒作为阴性对照，结果为以农杆菌为模板，PCR 扩增出 447 bp 左右的片段，以空载体质粒为模板没有出现相应的条带（图 3-30）。

第3章 辣椒对低温胁迫的响应与其低温抗性相关基因的克隆和功能分析

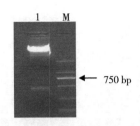

图 3-29 病毒载体 pTRV2-*CaNAC2* 双酶切检测
泳道 1：pTRV2-*CaNAC2* 双酶切产物；M：2*Kb* Marker

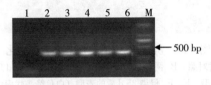

图 3-30 病毒沉默载体 pTRV2-*CaNAC2* 的菌液 PCR 检测
泳道 1：以空载体质粒 pTRV2 为模板的 PCR 产物；
泳道 2～6：转化 pTRV2-*CaNAC2* 质粒的菌液 PCR 产物；M. 2*Kb* Marker

(3) *CaNAC2* 沉默效率分析

①pTRV2-*PDS* 沉默结果与检测。以子叶完全展开，第 1 片真叶刚长出的辣椒植株为试验材料，采用注射法将含有构建病毒载体的农杆菌注入辣椒叶片，注射不含目的片段的 pTRV2 空载体的农杆菌作为阴性对照。子叶注射侵染后针孔处留下明显的干枯状痕迹（图 3-31C），接种 15 d 后，含有 pTRV2-*PDS* 病毒载体的辣椒栽培种 P70 叶柄处出现黄化现象（图 3-31C）；20 d 后，侵染 pTRV2-*PDS* 病毒载体的植株大部分叶片出现光漂白现象，新长出的真叶完全白化（图 3-31D）；继续生长到 25 d，大部分叶片和茎段白化现象更明显（图 3-31E、F）。而阴性对照植株没有出现白化症状（图 3-31A）。由此可见，VIGS 技术可以使辣椒栽培种 P70 有效地发生基因沉默。另外，根据 *PDS* 沉默出现的白化症状，我们可以预测辣椒栽培种 P70 发生病毒诱导基因沉默的大概时间。

②pTRV2-*CaNAC2* 沉默效率分析。接种 30 d 后，调查单株第 5、6 叶位 *CaNAC2* 转录水平。在病毒诱导基因沉默系统中，插

图 3-31 沉默 *PDS* 的辣椒 P70 表型变化

A. 未注射对照 B. 阴性对照侵染后表型 C. 侵染 15 d 后的表型
D. 侵染 20 d 后的表型 E、F. 侵染 25 d 后的表型（白色箭头代表侵染后留下的痕迹）

入的外源基因会随病毒载体在植物各组织中大量转录，因此，利用该片段为目标检查基因的沉默效率会受到影响。我们在病毒载体包含目的基因的外围区域合成一对特异引物（F: 5′-AAACGCTTGTCTTCCATAGTG-3′; R: 5′-CCACATCATCCTCTGCCTC-3′）。与阴性对照相比，pTRV2-*CaNAC2* 感染植株 *CaNAC2* 表达量下降 72%（图 3-32）。

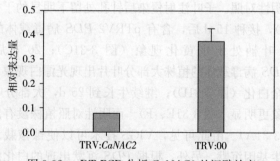

图 3-32 qRT-PCR 分析 *CaNAC2* 的沉默效率

（4）*CaNAC2* 沉默对辣椒应答低温和 ABA 处理的抗性分析

外源 ABA 喷施 72 h 后，对辣椒叶片进行低温处理 24 h，发生

CaNAC2 沉默的植株丙二醛含量高于对照,但差异不显著(图 3-33A);然而,*CaNAC2* 沉默的植株过氧化氢含量显著高于对照(图 3-33B),结果表明 *CaNAC2* 沉默使辣椒植株耐寒性减弱,这与第 3 章研究的结果一致:ABA 可能诱导 *CaNAC2* 来提高辣椒的耐寒性。发生 *CaNAC2* 沉默的植株直接进行低温处理,丙二醛含量显著高于对照(图 3-33C);然而,沉默植株过氧化氢含量很低,显著低于对照(图 3-33D),可能与 *CaNAC2* 沉默的植株过氧化氢参与信号转导有关。

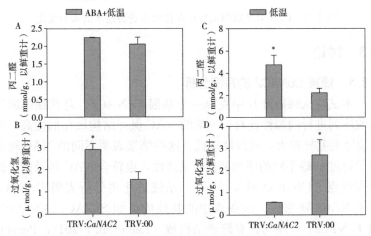

图 3-33 外源 ABA 和低温对 *CaNAC2* 沉默的辣椒丙二醛和过氧化氢含量的影响

(5)*CaNAC2* 沉默对辣椒响应盐胁迫和渗透胁迫的抗性分析 靠近发生基因沉默叶片的上部叶片(第 5 或 6 叶位)进行 NaCl 和甘露醇胁迫处理。离体圆形叶片漂浮在 300 mmol/L NaCl 模拟的盐胁迫 3 d 之后,*CaNAC2* 沉默的植株叶片均保持绿色,而阴性对照植株叶片出现光漂白现象,特别是叶脉处和叶片边缘(图 3-34)。因此,我们推测盐胁迫下辣椒 *CaNAC2* 的缺失与植株保绿有关。300 mmol/L 甘露醇处理下,*CaNAC2* 沉默的植株和阴性对照基本没有明显差异(图 3-34)。

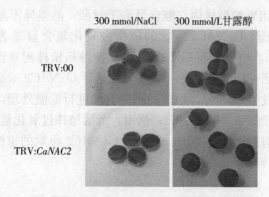

图 3-34 CaNAC2 沉默的辣椒在盐和渗透胁迫下的表型变化

4.3 讨论

4.3.1 辣椒 CaNAC2 的特征分析

本试验从辣椒叶片中克隆一个基因 CaNAC2，对其编码的氨基酸序列进行同源性比对，N 端的 NAC 保守结构域相似性高，而 C 端序列差异较大，同源性较低。这些结果表明不同的 NAC 转录因子通过 C 端不同的序列调控转录活性，也符合 NAC 转录因子的结构特点（Olsen et al.，2005）。系统进化树分析表明 CaNAC2 属于 NAC2 亚家族。该亚族中的其他成员如 SINAC、ANAC053 和 TaNTL5，C 端均含有跨膜结构域（Lee et al.，2012；Puranik et al.，2011；Tang et al.，2012）。然而采用 TMpred 跨膜结构分析软件分析 CaNAC2 的发现 C 端不含有跨膜结构域（数据未给出）。这些结果表明属于同一亚族的 NAC 之间 C 端序列差异很大，并暗示这些基因之间的功能存在多样性。一些与膜结合的 NAC 蛋白含有 335~652 个氨基酸，因为 C 端跨膜结构域的延伸，其分子量要比核 NAC 蛋白大（Morishita et al.，2009）。

序列分析表明 CaNAC2 含有一个核定位信号区。基因枪轰击法证明 CaNAC2 定位于细胞核。同时也证实了 CaNAC2 C 端不含有跨膜结构。因为该基因核定位的特征不同于 NAC2 亚族中的 NTL 转录因子（C 端含有跨膜序列的 NAC 基因）定位：NTLs 常定位于细胞质和细胞核（Lee et al., 2012；Puranik et al., 2011；Tang et al., 2012）。NTL 转录因子通常与胞内膜结合以休眠的形式存在，当外界环境因子刺激时通过蛋白水解活化（Kim et al., 2007；Seo et al., 2008）；活化的 NTL 进入细胞核，调节参与非生物逆境信号转导的基因表达如干旱、低温和高盐（Kim et al., 2006, 2008；Seo et al., 2010）。大量研究表明 NAC 转录因子 C 端具有转录激活活性，富含几种氨基酸的简单重复序列，如 Ser、Trp、Pro 和 Glu（Hegedus et al., 2003；Kikuchi et al., 2000；Souer et al., 1996；Xie et al. 1999），这是植物转录激活区域的共同特征（Liu et al., 1999；Olsen et al., 2005）。*CaNAC2* 编码的氨基酸 C 端转录激活区域也含有一个这样的特征。我们利用酵母功能互补实验进行转录活性分析，证明该基因具有转录激活特性（图 3-20）。

4.3.2 辣椒 *CaNAC2* 的抗逆性分析

组织表达特异性的转录因子在植物的生长和发育中起着重要的作用（He et al., 2005；Mitsuda et al., 2007；Yoo et al., 2007）。本试验 *CaNAC2* 在辣椒各个组织中均有表达，根和幼嫩种子中表达量较大，表明该基因可能在根系生长和种子成熟过程中起着重要的作用。*AtNAC2* 主要在根中表达，能够促进侧根发育（Xie et al., 2000）。种子发育和萌发期间诱导 *CarNAC1* 表达（Peng et al., 2010）。

植物 NAC 响应非生物逆境反应。有些 *NAC* 基因能够提高植物的抗逆性如拟南芥、水稻、大豆和其他作物。水稻 *OsNAC6* 在低温、高盐、干旱和 ABA 处理下上调表达（Nakashima et al., 2007；Ohnishi et al., 2005）。并且该基因在水稻上的过量表达能够提高作物的抗逆性（Nakashima et al., 2007）。油菜 *BnNACs* 受机械损伤、低温和干旱诱导表达（Hegedus et al., 2003）。大豆中有 9 个 *NAC* 基因在以下一个或几个胁迫处理中上调表达，如干旱、高盐、低温和 ABA 处理（Tran et al., 2009），其中 *GmNAC20* 的过量表达提高拟南芥植株的抗逆性，并通过调节与生长素信号相关的基因表达促进侧根生长（Hao et al., 2011）。小麦6个 *NAC* 基因（*TaNAC2a*、*TaNAC4a*、*TaNAC6*、*TaNAC7*、*TaNAC13* 和 *TaNTL5*）在不同的逆境胁迫下表达模式不同：失水和高盐胁迫诱导 *TaNAC2a* 表达，而抑制 *TaNTL5* 表达；ABA 和低温对 *TaNTL5* 表达没有影响（Tang et al., 2012）。SA 在生物胁迫信号转导中的作用研究很多，目前发现也参与非生物胁迫反应。外源 SA 能够缓解逆境胁迫对植物造成的伤害，如低温、高盐和重金属等（Horváth et al., 2007）。辣椒 *CaNAC1* 受 SA 诱导，并参与 SA 介导的病原菌信号反应（Oh et al., 2005）。本试验中，SA 抑制 *CaNAC2* 表达，表明 SA 可能介导 *CaNAC2* 响应重金属胁迫反应。

在干旱胁迫下，*NTL4* 突变体叶片过氧化氢含量低于野生型，过量表达株系通过正调节活性氧合成基因 Atrboh 引起过氧化氢的过量积累，导致电导率的增加（Lee et al., 2012）。*RD26* 受过氧化氢诱导表达，过量表达 *RD26* 的植株上调一些防御相关基因如谷胱甘肽转移酶和 aldo/keto 还原酶家族基因的表达（Fujita et al., 2004），他们在活性氧的解毒中起着重要的作用（Oberschall et al., 2000）。我们以前的报道：ABA、低温和两者结合处理诱导 *CaNAC2* 表达，ABA＋低温处理对其表达上调的程度高于低温处理。这些结果表明 *CaNAC2* 可能通过依靠 ABA 信号转导响应低温反应。为了进一步分析 *CaNAC2* 对辣椒耐寒性的影响，我们采用病毒诱导沉默技术对其进行敲除，然后研究参与 ABA 和低温处理下的反应。外源 ABA 喷施后进行低温处理和喷施清水后直接进行相应的低温处理，发生基因沉默的植株丙二醛含量高于对照，表明 *CaNAC2* 沉默后，辣椒膜脂过氧化的程度加剧，植株的耐寒性减弱，而过量表达此基因是否能够增加辣椒的耐寒性需要进一步研究。ABA＋低温处理发生基因沉默的植株过氧化氢含量高于对照；然而，仅遭受低温处理的基因沉默植株过氧化氢含量很低，这可能与过氧化氢参与信号转导有关。因此我们推测 *CaNAC2* 与过氧化氢的代谢有关，但该基因如何调节活性氧的代谢有待于进一步研究。

NAC 转录因子通过依赖 ABA 和不依赖 ABA 途径调节非生物逆境反应 (Fujita et al., 2004; Tran et al., 2004)。本试验 ABA 逐渐诱导辣椒 *CaNAC2* 表达；在高盐胁迫反应中该基因也上调表达，表明 *CaNAC2* 响应高盐胁迫反应可能是通过 ABA 信号介导，这与以上提及的水稻 *OsNAC6* 和大豆 *GmNAC20* 表达响应逆境反应的变化一致。甘露醇处理抑制 *CaNAC2* 表达，表现出与 ABA 处理相反的表达。另外，高盐和甘露醇处理影响基因的表达模式不同于以前的报道：很多失水诱导的基因表达在高盐胁迫下也表现出相近的结果 (Tran et al., 2009; Zhou et al., 2008)。另外，我们研究了失去 *CaNAC2* 功能的植株响应高盐和甘露醇胁迫反应的变化。与未发生基因沉默的对照相比，在高盐胁迫下，发生基因沉默的离体叶片保持较绿，也就是衰老发生延迟。此结果表明该基因的功能缺失与植株保绿有关。*CaNAC2* 的超量表达是否出现相应的叶片衰老加速很值得进一步研究。有类似的报道：NTL4 突变体通过依靠 ABA 信号途径延迟干旱诱导的叶片衰老，在干旱胁迫下超量表达株系叶片衰老加速，耐旱性降低 (Lee et al., 2012); *AtNAP* 与叶片的衰老有关，超量表达 *AtNAP* 引起叶片衰老 (Guo and Gan, 2006)。而经甘露醇处理，发生基因沉默的离体叶片与对照相比，表型基本没有发生变化。总之，*CaNAC2* 沉默的植株对低温和高盐胁迫反应的表现不同，表明该基因的功能可能与逆境胁迫种类有关。水稻 *SNAC1* 的过量表达株系在高盐和干旱胁迫反应时抗性增强，但在遭受低温胁迫时抗性没有显著变化 (Hu et al., 2006)。

5 辣椒 CaMBF 转录辅激活因子的功能分析

逆境对全球的农业产量造成很大的损失。为了适应各种逆境如低温、高盐或干旱，植物体内发生了由复杂的转录因子网络和其他调节基因如调控多种防御反应的相关酶和蛋白介导的一系列变化。转录辅激活因子与转录因子的活性有关，能够增强转录因子和顺式作用元件的结合。MBF1 转录辅激活因子通过结合通用转录因子 TATA 元件结合蛋白（TBP）和基因特异的转录因子如基本的 bZIP 蛋白家族，从而介导基因的转录激活（Kabe et al.，1999；Takemaru et al.，1997，1998）。MBF1 是高度保守的转录辅激活因子，参与各种代谢过程的调节，如内表皮细胞分化、激素调节的脂类代谢（Brendel et al.，2002；Liu et al.，2003；Takemaru et al.，1997，1998）。MBF1 的超量表达引起了很多编码逆境反应相关的转录因子和信号转导基因的积累，如 DREB2A、WRKY、锌指蛋白 Zat7 和 Zat12、APX2 和 PRs 蛋白等（Suzuki et al.，2005，2011）。

模式植物拟南芥有 3 个不同的基因（*AtMBF1a*、*AtMBF1b* 和 *AtMBF1c*）编码 MBF1 蛋白，功能分析表明这 3 种同系物存在明显不同的表达模式，作为转录辅激活因子在植物的抗逆性方面起着重要的作用（Tsuda and Yamazaki，2004）。*AtMBF1a* 受失水和高葡萄糖诱导表达，且在拟南芥上该基因的组成型超量表达提高植株的耐盐性、对葡萄糖的不敏感性和抗病性，且没有抑制植物的生长（Kim et al.，2007）。*AtMBF1c* 响应逆境胁迫时如高温、过氧化氢、失水和高盐，上调表达（Tsuda and Yamazaki，2004）。并且 *MBF1c* 的超量表达部分启动或扰乱乙烯信号转导途径，进而能够提高转基因拟南芥对高温和渗透胁迫的抵抗能力，然而转基因植株的耐寒性与野生型相似（Suzuki et al.，2005），并且在拟南芥和大豆上 *AtMBF1c* 的超量表达提高作物的产量（Suzuki et al.，2005，2011）。另外 *MBF1* 的功能与植物的生长变化有关，如植株大小、叶片细胞宽度（Hommel et al.，2008；Tojo et al.，2009），*AtMBF1s* 依靠负调节 ABA 抑制种子萌发（Mauro et al.，2012）。

总之，将 *MBF1* 导入拟南芥中过量表达，转基因株系提高抗逆性的同时并没有对植株的生长产生抑制，表明 *MBF1* 可以作为提高植株产量和品质的候选基因。然而目前关于辣椒 *MBF1* 对抗逆性的影响报道很少，因此，本试验旨在研究 *MBF1* 与辣椒抗逆性的关系，进一步揭示 *MBF1* 的功能。主要研究内容：①辣椒 *CaMBF* 的特征分析；②在辣椒不同组织和非生物逆境胁迫下的表

第3章 辣椒对低温胁迫的响应与其低温抗性相关基因的克隆和功能分析

达模式分析，如 SA、高盐、甘露醇和汞胁迫；③$CaMBF$ 转基因拟南芥的耐寒性分析。

5.1 材料与处理

5.1.1 主要材料及处理方法

本研究拟采用的辣椒材料为 P70 栽培种，由西北农林科技大学辣椒课题组提供。辣椒生长到 6～7 片真叶，进行 ABA、SA、低温、盐和干旱处理，处理方法同第 4 章描述，汞处理，材料浸泡在 300 μmol/L 汞溶液中。

5.1.2 辣椒 CaMBF 分离和序列分析

以 SSH 反交库 EST（登录号 JZ198811）为探针，按照第 3 章第 4 节（4.1.2）描述的同源克隆法获得基因全长。$CaMBF$ 全长引物为：

F：5′-GAAGAAAAAAAGCAATGAGT-GG-3′

R：5′-GCAGAAACGAATTTAGGATTTG-3′

利用 MEGA5 软件进行分子进化分析，具体方法见第 3 章第 4 节描述。

5.1.3 辣椒 CaMBF 在非生物胁迫下的表达模式分析

植株生长到 6～7 片真叶，进行 ABA、SA、低温、盐、渗透和汞处理。RNA 提取，单链 cDNA 合成，实时定量均参考第 3 章第 3 节。

5.1.4 CaMBF 超量表达载体的构建

将 $CaMBF$ 全长插入到植物过量表达载体 pVBG2307（Ahmed et al., 2012），构建流程如图 3-35 所示：

所用引物为：

CaMBF-F：5′-GCAGAAACGAATTTAGGATTTG-3′

CaMBF-R：5′-CGCGGATCCGCAGAAACGAATTTAGGATTTG-3′（下游引物添加 Bam HI 酶切位点和保护碱基，下划线为 Bam HI 酶切位点）。

所用限制性内切酶为：Xba I 和 Bam HI。

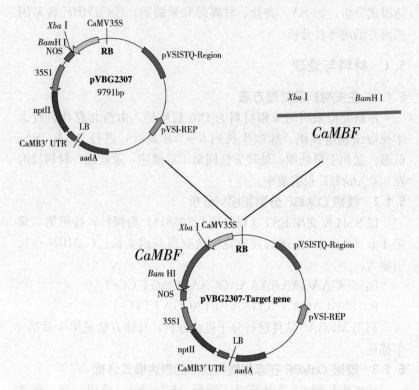

图 3-35 超量表达载体 pVBG2307-*CaMBF* 的构建示意图

按照 4.1.3.1 节描述的方法进行酶切，连接和转化：用 *Xba* I 和 *Bam* HI 限制性内切酶酶切 pMD19-T-*CaMBF* 载体和过量表达载体 pVBG2307。T4-DNA ligase 连接目的片段与酶切后的空载体，重组质粒 pVBG2307-*CaMBF* 转化大肠杆菌，在 50 mg/L K^+ 培养基筛选转化因子。挑取单菌落进行质粒提取，PCR 和双酶切检测载体构建正确后，表达载体 pVBG2307-*CaMBF* 转入农杆菌 GV3101，在10 mg/L Rif+50 mg/L Gen+50 mg/L K^+ 条件下筛选转化子。

5.1.5 辣椒 *CaMBF* 的拟南芥遗传转化及检测

（1）蘸花法侵染拟南芥　采用农杆菌介导的蘸花法（Clough

and Bent,1998) 对野生型拟南芥 (Col-0) 进行转化。具体步骤:

①拟南芥野生型 Col-0 营养生长期光周期:12 h/12 h (白天/晚上);生殖生长期光周期:16 h/8 h (白天/晚上)。温度设为 22℃/18℃ (昼夜),相对湿度 70%。

②拟南芥顶生花序生长 2 cm 左右时,从基部去除,诱导侧生花序的生长。

③取 -80℃ 冻存的农杆菌 GV3101,含有表达载体质粒,在含有 Rif (10 mg/L) +Gen (50 mg/L) +K$^+$ (50 mg/L) 抗生素的平板上划板,然后挑取单菌落进行 28℃ 摇菌培养至 OD_{600} 为 0.6~1.0。

④4℃,5 000 r/min,离心 2 min,收集菌体,将农杆菌沉淀悬浮于新配的转化缓冲液中 (1/2 MS 培养基,0.05% Silwet L-77),至终浓度 OD_{600} 为 0.5~0.8。

⑤转化前一天浇水,转化时去除果荚和正在盛开的花朵,将拟南芥的花序浸入转化缓冲液中 10~30 s,取出平放植株,覆盖保鲜膜,22℃ 遮光 12~24 h。

⑥揭去保鲜膜,将植株正常竖直培养,待果荚成熟后收获种子 T_1 代。

⑦为提高转化效率,可多次蘸花处理。间隔 5~7 d 进行 2 次蘸花处理。

(2) 拟南芥种子的消毒 首先将转基因拟南芥种子放入 75% 乙醇 45 s,倒出。然后用 1/8 次氯酸钠 (体积比) 消毒 20 min,无菌水冲洗 4 次;再用 0.05% 琼脂水悬浮均匀涂在筛选培养基 (50 mg/L K$^+$) 上。种子 4℃ 春化 3~4 d。然后转移至正常条件生长。种子萌发后,子叶逐渐黄化的是非转基因植株,子叶深绿色正常生长的是转基因植株。此转基因植株生长获得 T_2 代种子,进行单株收种,纯系 T_2 代种子进行以下试验处理。

(3) 转基因拟南芥检测

①T_1 代转基因拟南芥的 PCR 检测。经 K$^+$ 抗生素筛出的抗性苗,需进一步进行 PCR 检测鉴定。以抗性苗的 DNA 为模板,分

别对表达载体上的基因片段和 *NPTII* 进行 PCR 检测，拟南芥野生型 Col-0 为阴性对照，连有基因片段的载体质粒为阳性对照。

所用引物：

NPT II-F：5′-AGACAATCGGCTGCTCTGAT-3′
NPT II-R：5′-TCATTTCGAACCCCAGAGTC-3′
CaMBF-RT-F：5′-TACATCTTTGGACACCAGGAA-3′
CaMBF-RT-R：5′-GCAGAAACGAATTTAGGATTTG-3′

②T_2 代转基因拟南芥的 cDNA 水平检测。T_2 代转基因植株幼苗提取总 RNA，反转录成 cDNA，以拟南芥 *elF4A* 为内参 (Gutierrez et al.，2008)，半定量分析 *CaMBF* 的表达量。RNA 提取，单链 cDNA 合成均参考第 3 章第 3 节。

所用引物：

*elF*4A-F：5′-TGACCACACAGTCTCTGCAA-3′
*elF*4A-R：5′-ACCAGGGAGACTTGTTGGAC-3′
CaMBF 检测引物同 *CaMBF*-RT。

（4）CTAB 法提取转基因植株基因组 DNA

5.1.6　*CaMBF* 超量表达拟南芥的表型分析与功能鉴定

（1）转基因拟南芥低温胁迫分析　2 周苗龄的 T_2 代拟南芥进行低温 4℃处理 48 h，观察表型变化。并分别在不同的低温处理时间（0、2、6、24 和 48 h）进行取样，液氮速冻，−80℃保存备用。试验进行 3 次重复。

（2）转基因拟南芥的 cDNA 合成和实时定量分析　总 RNA 提取，单链 cDNA 合成，实时定量均参考第 3 章第 3 节。所用引物如表 3-12。

表 3-12　基因引物序列

基因	登录号	上游和下游引物 5′→3′
AtRD22	At5g25610	F：GCGTTGGCAGCGGAAAA R：GCGTTAGGATCGTCGTGG
		F：CTGATCCCACCAAAGAAGAAACT

(续)

基因	登录号	上游和下游引物 $5'\rightarrow 3'$
AtRD29A	At5g52310	R：AAGCCCATCGGAGAATTC F：AAGAAGAACATGGCGTCTTACC
AtRAB18	AT5G66400	R：GTTCCAAAGCCTTCAGTCCC F：CCAGCGAAATGGGGAAAC
AtERD15	At2g41430	R：ACAAAGGTACAGTGGTGGC F：AAATGTCAGAGACCAACAAGAA
AtKIN1	At5g15960.1	R：CTACTTGTTCAGGCCGGTCTT F：TCGGTTTATGCAGCATTGGA
AtPR2	At3g57260	R：TAACAACATACTACACGCTGAAAG F：ACAGAAGTTGTGCGAGAAGC
AtPDF1.2	At5g44420	R：CACGATTTAGCACCAAAGATT

5.2 结果与分析

5.2.1 辣椒 CaMBF 全长 cDNA 的分离

以 SSH 反交库中筛选得到的 MBF 同源 EST（R028）为探针，P70 栽培种 cDNA 为模板，参考第 3 章第 4 节（4.1.2）描述的同源克隆法获得基因全长 cDNA 序列，命名为 CaMBF。CaMBF cDNA 全长为 648 bp（GenBank 登录号：JX402927），开放阅读框为 420 bp，编码 140 个氨基酸，分子量为 15.3 KD，等电点为 9.86。基因测序获得的 cDNA 全长及其氨基酸序列如图 3-36。

为了分析 CaMBF 编码的氨基酸结构特点，利用 DNAMAN 软件进行与其他物种的 MBF 蛋白进行序列比较，CaMBF 编码的氨基酸具有高同源特点（图 3-37）。辣椒 CaMBF 与番茄 MBF1（登录号：NP_001234341.1）、马铃薯 StMBF1（登录号：AAF81108.1）和拟南芥 MBF1b（登录号：NP_191427.1）同源性高，分别可达 94.37%、95.07%和 79.58%。

利用 MEGA5 软件进行 CaMBF 与其他已报道的 MBF 蛋白进化分析，结果如图 3-38 所示。根据 Kenichi Tsuda 和 Kenichi Yamazaki（2004）的分类，CaMBF 属于亚家族 I。

```
1 GAAGAAAAAAAGCA
15 ATGAGTGGAATATCACAAGACTGGGAGCCGGTGGTAATTCGTAAGAAGGCGCCGACCGCCGCC
    M  S  G  I  S  Q  D  W  E  P  V  V  I  R  K  K  A  P  T  A  A
79 GCTCGCAAAGACGAAAAAGCCGTCAACGCCGCACGTCGCGCCGGCGCCGAGATCGAAACCGTC
    A  R  K  D  E  K  A  V  N  A  A  R  R  A  G  A  E  I  E  T  V
142 AAAAAGTCTAATGCTGGTTCAAACAGGGCTGCTTCAAGTAGTACATCTTTGGACACCAGGAAA
    K  K  S  N  A  G  S  N  R  A  A  S  S  S  T  S  L  D  T  R  K
205 CTTGATGAAGACACTGAGAATTTGTCTCATGAAAAGGTACCAACTGAACTGAAGAAAGCCATC
    L  D  E  D  T  E  N  L  S  H  E  K  V  P  T  E  L  K  K  A  I
268 ATGCAAGCACGACAAGACAAGAAGCTCACTCAGGCTCAACTTGCTCAATTGATAAATGAGAAG
    M  Q  A  R  Q  D  K  K  L  T  Q  A  Q  L  A  Q  L  I  N  E  K
330 CCACAGATCATCCAAGAATACGAGTCCGGTAAGGCAATTCCAAACCAGCAGATAATCTCTAAA
    P  Q  I  I  Q  E  Y  E  S  G  K  A  I  P  N  Q  Q  I  I  S  K
393 CTGGAAAGAGCTCTTGGAGCGAAACTTCGAGGAAAGAAATGAAAGTGTGATTGCACAAGT
    L  E  R  A  L  G  A  K  L  R  G  K  K  *
453 TAAGCCTTATGGATTGTCACAGTCTGTTTGAATCCAAAATCCAGTCGTGTACG
506 CAAATTTTTCTTTAAACATCTGTGTGTTTTGCTTATCGTTATGTCTGGATTCATG
561 ACTACGGTTTTCTTTGTTGTATAGCTGTTTCTTCTGTGGGAAATCATCGTAAGA
615 AATCATAAACTTCAAATCCTAAATTCGTTTCTGC
```

图 3-36 辣椒转录辅激活因子 *CaMBF* cDNA 及其推导的氨基酸序列

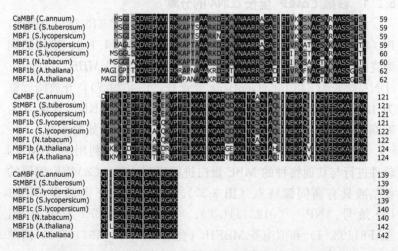

图 3-37 辣椒 CaMBF 与其他 MBF 蛋白的氨基酸序列比对

其他物种的 MBF 蛋白为：马铃薯 StMBF1（AAF81108.1）；番茄 MBF1（NP_001234341.1）、MBF1b（XP_004251896.1）、MBF1c（ABG29114.1）；烟草 MBF1（BAB88859.1）和拟南芥 MBF1A（NP_565981.1）、MBF1b（NP_191427.1）

第3章 辣椒对低温胁迫的响应与其低温抗性相关基因的克隆和功能分析

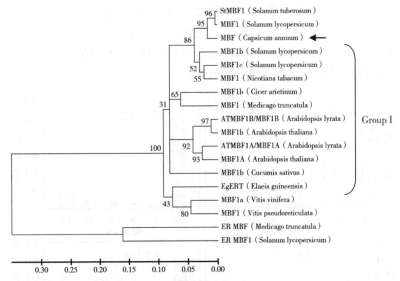

图 3-38 辣椒 CaMBF 与其他物种的 MBF 蛋白的进化树分析

5.2.2 辣椒 *CaMBF* 的表达模式分析

(1) *CaMBF* 在辣椒不同组织中的表达分析　qRT-PCR 结果表明 *CaMBF* 在辣椒不同组织中均有表达（图 3-39），在花和未成熟的种子中表达量较高，根中基本没有表达。

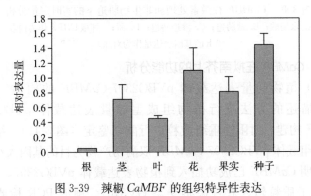

图 3-39 辣椒 *CaMBF* 的组织特异性表达

(2) *CaMBF* 响应逆境胁迫反应的表达分析　SA 处理 12 h 期间抑制其表达，24 h 逐渐恢复到正常水平（图 3-40A）。低温诱导

CaMBF 表达，表达模式变化很大：1 h 迅速增加，6 h 达到峰值，12 h 和 24 h 突然下降，48 h 又升到最大；在 NaCl 模拟的高盐胁迫、甘露醇模拟的高渗胁迫和重金属汞处理下，*CaMBF* 表达在整个处理期间均明显受到抑制（图 3-40B、C、D）。

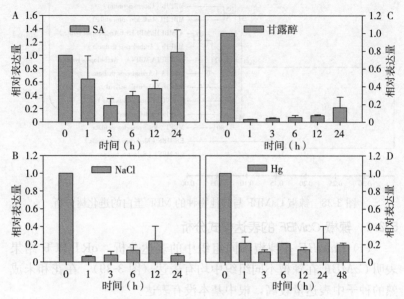

图 3-40 *CaMBF* 在激素处理和非生物胁迫下的实时定量分析
A. 水杨酸；B. 盐胁迫；C. 渗透胁迫；D. 汞；（辣椒 *UBI-3* 作为内参；0 h 的基因表达量作为对照）

5.2.3 *CaMBF* 在拟南芥上的功能分析

（1）植物超量表达载体 pVBG2307-*CaMBF* 的构建　按照 5.1.4 描述的方法进行植物组成型超量表达载体 pVBG2307-*CaMBF* 构建。对阳性重组质粒进行酶切鉴定（图 3-41），结果表明：重组载体 pVBG2307-*CaMBF* 双酶切产物为目的基因大小的条带，表明 *CaMBF* 已成功插入到植物表达载体 pVBG2307。然后，将转化子质粒转入农杆菌感受态 GV3101。菌液 PCR 检测转入 pVBG2307-*CaMBF* 的菌株，以空载体 pVBG2307：00 质粒作为阴性对照，结果证明以农杆菌为模板，PCR 扩增出与目的片段大小

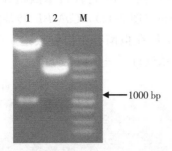

图 3-41 植物超量表达载体 pVBG2307-*CaMBF* 双酶切检测

泳道 1：pVBG2307-*CaMBF* 双酶切产物；泳道 2：pVBG2307-*CaMBF* 质粒；M：Trans2K Plus DNA Marker

一致的片段，空载体质粒为模板没有出现条带（图 3-42）。

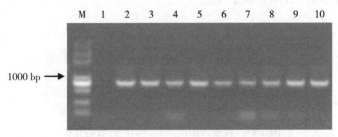

图 3-42 超量表达载体 pVBG2307-*CaMBF* 的菌液 PCR 检测

泳道 1：以空载体质粒 pVBG2307 为模板的 PCR；泳道 2-10：转化 pVBG2307-CaMBF 质粒的菌液 PCR；M. Trans2K Plus DNA Marker

（2）*CaMBF* 转基因拟南芥的获得 本研究采用农杆菌介导的蘸花法进行拟南芥野生型 Col-0 遗传转化，使 *CaMBF* 在拟南芥中组成型超量表达，其上收获的 T_1 种子在 50 mg/L K^+ 培养基上进行筛选，获得 T_1 代阳性拟南芥植株，再对其进行 DNA 和 cDNA 水平检测。结果表明，以转基因拟南芥 DNA 为模板扩增出大小一致、特异的目的条带，对照野生型没有扩增出相应的条带（图 3-43），表明 *CaMBF* 整合到拟南芥的基因组。经过 *CaMBF* 和 *NPTII* 的 PCR 检测，共获得 56 株转基因拟南芥。而且，我们进一步以野生型 Col-0 拟南芥和 T_1 代转基因植株的 cDNA 为模板，进行半定量 PCR 分析（图 3-44），CaMBF 转基因植株扩增出相应的条带，表明 CaMBF 在拟南芥中得到表达，选取表达量较高的 T5-12 和 T15-21 株系进行下一步功能分析。

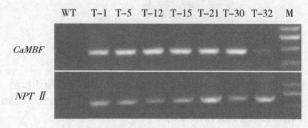

图 3-43　转基因拟南芥植株 CaMBF 和 NPTII 的 PCR 检测
WT：野生型；转基因株系：T-1、T-5、T-12、T-15、T-21、T-30 和 T-32；M：Trans2K Plus DNA 分子标记

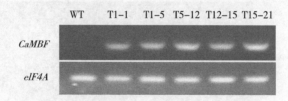

图 3-44　转基因拟南芥 CaMBF 的表达分析
WT：野生型；转基因株系：T1-1、T1-5、T5-12、T12-15 和 T15-21

（3）CaMBF 转基因拟南芥植株的表型分析　为了分析辣椒 CaMBF 在拟南芥上的功能，将带有 35 S 的 CaMV 组成型启动子超量表达载体导入拟南芥，分析 CaMBF 在非生物逆境胁迫下的表达变化。如图 3-45 所示，CaMBF 在转基因拟南芥植株中大量表达，而在野生型幼苗中基本没有表达。与对照相比，在盐、低温胁迫和 ABA 处理下转基因植株 CaMBF 表达量下降，表明 CaMBF 负调节非生物逆境反应。

在正常生长条件下，CaMBF 转基因拟南芥植株在生长和发育上与野生型表现相似。营养生长初期转基因植株和野生型长势一致，而在花絮出现期间，转基因植株生长较大，莲座叶的叶长和叶

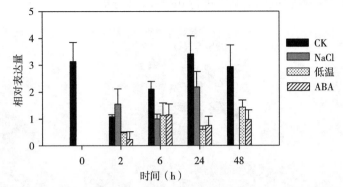

图 3-45　转基因株系 CaMBF 响应非生物逆境胁迫的表达变化

宽分别比对照多 70% 和 60%（图 3-46）。这与 AtMBF1c 转基因拟南芥表型一致（Suzuki et al.，2005）。

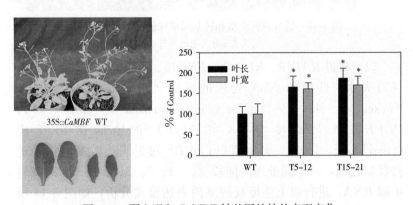

图 3-46　野生型和 CaMBF 转基因植株的表型变化

WT：野生型；T5-12 和 T15-21：转基因株系

（4）辣椒 CaMBF 对拟南芥的耐寒性分析　为了分析 CaMBF 转基因植株的耐寒性，对苗龄和大小一致的转基因植株和野生型进行低温 4℃ 处理，结果表明：转基因植株在低温 6 h 时出现萎蔫，24 h 加重，而野生型低温 48 h 才出现萎蔫（图 3-47），表明 CaMBF 的超量表达降低了拟南芥的耐寒性。我们推测拟南芥 CaMBF 的超量表达下调了一些逆境反应基因。为了验证这一假设，我们对一些逆境标记基因进行 qRT-PCR 分析。

图 3-47　低温对野生型和转基因植株幼苗生长的影响
箭头表示叶片出现萎蔫

以前报道 *RAB18*、*RD29A*、*ERD15*、*KIN1* 和 *RD22* 参与植株对失水、低温和 ABA 处理的反应（Kiyosue et al.，1994；Kurkela et al.，1992；Yamaguchi-Shinozaki et al.，1993）。*PDF1.2* 和 *PR2* 参与 SA 和 JA/ET 介导的防御反应（Glazebrook，2005）。野生型和 *CaMBF* 过量表达转基因株系进行低温处理，对不同处理时间段（0、2、6、24 和 48 h）的材料提取 RNA，进行以上逆境反应基因表达模式分析，如图 3-48 所示，拟南芥 *eIF4A* 为内参，逆境相关基因表达量相对于野生型 0 h 的表达。

在正常生长条件下，*CaMBF* 超量表达转基因植株 *RD29A*、*RAB18*、*ERD15*、*KIN1* 基因表达整体上高于对照（图 3-48A、B、D 和 E）。转基因植株 *RD22* 和 *PR2* 表达在 0 h 和 48 h 时高于对照，其他时间较低（图 3-48C 和 G）。与常温相比，低温 24 h 期间 *CaMBF* 转基因株系和野生型 *RD29A*、*ERD15* 和 *KIN1* 表达显著增加（图 3-48A、D 和 E），表明低温强烈诱导 *RD29A*、*ERD15*

第3章 辣椒对低温胁迫的响应与其低温抗性相关基因的克隆和功能分析

和 *KIN1* 表达；在低温胁迫条件下，转基因株系 *RD29A*、*ERD15*（48 h 除外）和 *KIN1* 表达量低于野生型，表明拟南芥 *CaMBF* 的超量表达降低了低温对 *RD29A*、*ERD15* 和 *KIN1* 的诱导水平。与常温相比，低温显著抑制转基因株系和野生型 *RAB18* 和 *RD22* 表达（图 3-48B 和 C）；在低温 48 h 期间，转基因植株 *RAB18* 表达与野生型相似（48 h 除外），*RD22* 表达量很低，低于对照，表明拟南芥 *CaMBF* 的超量表达加重低温对 *RD22* 的抑制。与常温相比，转基因和野生型植株 *PDF1.2*（48 h 除外）和 *PR2* 在低温期

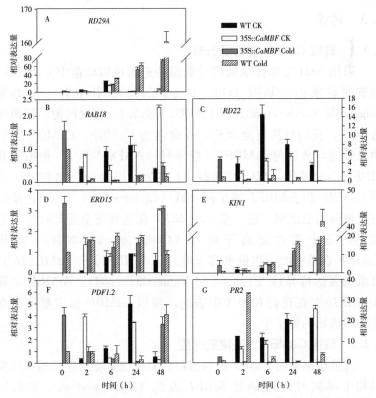

图 3-48 在低温胁迫下，野生型和 *CaMBF* 超量表达转基因植株逆境反应基因的表达分析

拟南芥 *elF4A* 作为内参，拟南芥野生型 0 h 的基因表达量作为对照

间下调表达（图 3-48F 和 G）；低温胁迫下，转基因植株 *PDF1.2* 和 *PR2* 表达量低于野生型。

总之，在低温胁迫条件下，*CaMBF* 超量表达转基因植株加剧低温对 *RD22*、*PDF1.2* 和 *PR2* 表达的抑制，同样转基因植株也抑制低温对 *RD29A*、*ERD15* 和 *KIN1* 的诱导表达，表明拟南芥 *CaMBF* 的超量表达负调节 *RD29A*、*ERD15*、*KIN1*、*RD22*、*PDF1.2* 和 *PR2* 的表达；转基因植株 *RAB18* 表达与野生型相似。

5.3 讨论

5.3.1 辣椒 CaMBF 的特征分析

采用 SSH 文库技术从遭受低温胁迫的辣椒幼苗中分离 ABA 调节的耐寒相关基因（Guo et al.，2013）。其中有一个差异 unigene 与 *S. tuberosum* MBF1（EU294363.1）高度同源，同源率为 80%。我们对其克隆全长，并命名为 *CaMBF*。*CaMBF* 编码的氨基酸与马铃薯 *StMBF1*（登录号：AAF81108.1）和拟南芥 *MBF1b*（登录号：NP_191427.1）的同源率达 95% 和 80%，表明 *CaMBF* 和 *StMBF1* 属于同一族（Godoy et al.，2001）；系统进化树分析也证明了这一点。*StMBF1* 在茎和完全展开叶中均有表达，茎中表达量高于叶片（Godoy et al.，2001）；辣椒 *CaMBF* 在茎中的表达量也高于叶片。*AtMBF1a* 和 *AtMBF1b* 具有组织表达特异性（Tsuda and Yamazaki，2004）。*AtMBF1a* 和 *AtMBF1b* 都在花药和种子中表达。辣椒 *CaMBF* 也主要分布在花和未成熟的种子中。

5.3.2 辣椒 CaMBF 对逆境的响应

大量研究表明 *MBF1* 参与植物的非生物逆境反应。烟草经高温和干旱同时处理诱导 *MBF1* 表达（Rizhsky et al.，2002）；ABA 对拟南芥 *AtMBF1a* 和 *AtMBF1b* 表达没有影响，而显著诱导 *AtMBF1c* 表达（Tsuda et al.，2004a）；马铃薯 *StMBF1* 受机械损伤、病菌侵染和 SA 强烈诱导（Godoy et al.，2001）。而辣

椒 *CaMBF* 受 SA 处理抑制表达。盐胁迫诱导拟南芥 *AtMBF1a* 和 *AtMBF1b* 表达（Kim et al.，2006），干旱和高温对 *AtMBF1a* 和 *AtMBF1b* 表达没有影响（Suzuki et al.，2005）。在高盐、甘露醇模拟的渗透胁迫和重金属汞处理下，辣椒 *CaMBF* 表达在整个处理期间均明显受到抑制。这些结果表明辣椒 *CaMBF* 在响应逆境反应时下调表达，如盐胁迫、渗透胁迫。这与拟南芥 *AtMBF1c* 对高盐和渗透胁迫的反应不同（Tsuda and Yamazaki，2004）。低温诱导辣椒 *CaMBF* 表达，这与以前报道的低温对 *MBF1* 表达影响不同：番茄 *LeMBF1* 在遭受低温胁迫 24 h 期间，表达量基本没有明显变化（刘阳，2007）。

5.3.3 辣椒 *CaMBF* 的耐寒性分析

在拟南芥野生型 Col-0 中，组成型超量表达辣椒 *CaMBF* 转基因植株较大，莲座叶的叶长和叶宽分别比对照多 70% 和 60%。这与 *AtMBF1c* 超量表达转基因拟南芥表型变化一致（Suzuki et al.，2005），转基因植株表型比对照大 20%；而抑制 *AtMBF1c* 表达的转基因植株茎伸长明显迟缓，植株矮小，叶柄和叶片长度显著减小，而且转基因植株的种子萌发率大量降低（Hommel et al.，2008）。*AtMBF1a* 超量表达拟南芥与野生型相比，形态和生长没有变化（Kim et al.，2007）。

在低温胁迫条件下，*CaMBF* 超量表达株系在 6 h 时出现萎蔫，而野生型遭受低温 48 h 时才出现萎蔫。表明 *CaMBF* 超量表达可能降低拟南芥的耐寒性。这与以前报道的 *MBF1* 与逆境胁迫之间的关系不同。拟南芥 *AtMBF1a* 组成型超量表达提高植株的耐盐性和抗病性（Kim et al.，2007）。*AtMBF1c* 超量表达也增强对高温、渗透胁迫及高温和渗透同时处理的抗性，然而转基因植株的耐寒性与野生型相似（Suzuki et al.，2005）。*LeMBF1* 超量表达也提高转基因番茄的耐热能力（王达菲，2009）。

为了探索转基因植株出现的低温症状和逆境相关基因表达之间的关系，我们对 *CaMBF* 超量表达株系和野生型进行低温处理，分析一些逆境相关基因的表达。在正常生长条件下，转基因植株

RD29A、*RAB18*、*ERD15*、*KIN1* 表达整体高于对照，表明拟南芥 *CaMBF* 的超量表达正调节这些基因表达。*AtMBF1a* 超量表达转基因拟南芥上调 *ERD15* 表达，对 *RD29A* 和 *RD22* 表达没有影响（Kim et al.，2007）。转基因和野生型拟南芥 *RD29A*、*ERD15* 和 *KIN1* 受低温诱导表达，而转基因植株 *RD29A*、*ERD15* 和 *KIN1* 表达量低于野生型；表明 *CaMBF* 超量表达降低低温对 *RD29A*、*ERD15* 和 *KIN1* 的诱导水平。*RD29A* 通常被认为是抗逆性的标记基因（Cheong et al.，2010；Yang et al.，2010）。*RD29A* 的表达可以被用作衡量与抗逆性之间的标准（Ding et al.，2009；Kim and Kim，2006），比如比较野生型和超量表达株系上的一些标记基因如 *RD29A*、*RD22* 和 *COR15* 进行判断是否参与非生物逆境反应（Yang et al.，2010）。*CBL5* 的超量表达改变逆境标记基因的表达如 *RD29A*、*RD29B* 和 *KIN1*，从而被推断为正调节盐胁迫和干旱胁迫反应（Cheong et al.，2010）。目前报道在高盐胁迫下，拟南芥 *AtMBF1a* 超量表达可以促进 *RD29A*、*RD22*、*ERD15* 和 *KIN2* 等表达（Kim et al.，2007）。在转基因植株和野生型上，低温显著抑制 *RAB18* 和 *RD22* 表达。在低温 24 h 期间，转基因植株 *RD22* 表达量低于对照，表明 *CaMBF* 超量表达加重低温对 *RD22* 的抑制；本试验中，*CaMBF* 超量表达降低拟南芥的耐寒性，这与 *CaMBF* 超量表达下调一些逆境反应基因有关，如 *RD29A*、*ERD15*、*KIN1* 和 *RD22*。

PDF1.2 和 *PR2* 是特异的防御反应标记基因，分别与乙烯信号和 SA 信号转导有关（Brown et al.，2003；Gu et al.，2002）。*PRs* 基因表达上调不仅可以提高植物的抗病性，还参与植物其他方面的抗胁迫能力（Datta and Muthukrishnan，1999）。低温强烈抑制转基因和野生型植株 *PDF1.2*（48 h 除外）和 *PR2* 表达；低温胁迫下，转基因植株 *PDF1.2* 和 *PR2* 表达低于野生型，表明 *CaMBF* 超量表达加剧低温对 *PDF1.2* 和 *PR2* 的抑制，因此我们推测 *CaMBF* 扰乱了乙烯和 SA 信号转导。在遭受真菌侵染的拟南芥植株上，*AtMBF1a* 超量表达抑制 *PR2* 表达，而激

活 *PDF1.2* 表达（Kim et al.，2007）；*AtMBF1c* 超量表达增强 *PR2*、*PR3*、*HSPs* 等表达，部分激活或扰乱乙烯信号转导（Suzuki et al.，2005）。

6 利用 VIGS 技术研究辣椒 CaF-box 蛋白基因的功能

通常，逆境相关基因分为两类：一类是信号蛋白，传导信号和转录因子进行基因调节；第二类是基因产物如渗透调节物和抗氧化物等（Bhatnagr-Mathur et al.，2008；Wang et al.，2003）。培育抗逆品种主要是调节这两类逆境相关基因，然而利用传统育种方法提高植物的抗逆性非常困难（McKersie et al.，1999；Vinocur et al.，2005）。与改变单一的逆境相关基因相比，转录因子遗传转化是个更有效的方法，因为转录因子可以调节很多下游逆境相关基因（Bhatnagar-Mathur et al.，2008；Chinnusamy et al.，2005）。响应不同逆境的很多基因共用转录因子，因此调节这些转录因子的表达可能提高多种抗逆性。另外提高植物抗逆性的方法是蛋白调节子。调节蛋白质活性的一个重要调控机制是翻译后修饰（Henriques et al.，2009）。逆境相关信号蛋白和转录因子的翻译后修饰是调节逆境相关基因表达的一个重要方法。泛素介导的蛋白质降解是真核细胞翻译后修饰调控的一个重要途径。

泛素参与生物和非生物逆境反应（Dreher et al.，2007；Lyzenga et al.，2011），通过调节激素合成，识别信号蛋白如转录因子的丰度等可以快速和有效响应逆境反应。泛素介导的依赖于 26S 蛋白质酶体的蛋白质降解途径主要由 3 个连续的酶促反应组成。E1、E2 序列结构高度保守，容易被识别；E3 结构差异最大，数目最多。泛素连接酶 E3 携带着底物特异性的信息，决定蛋白降解途径的底物特异性。E3 有以下几种类型组成：HECT、APC、VBC-Cul2、Ring/U-box 和 SCF。SCF（Skp1、Cul1/Cdc53、

Roc1/Rbx1/Hrt1、F-box)是 E3 中最大、最具特点且被广泛研究的家族之一（Petroski et al., 2005; Schwechheimer et al., 2009）。F-box 蛋白在 N 端含有约 60 aa 高度保守的 F-box 结构域，结合 SKP1（Lechner et al., 2006），C 端与底物结合，从而介导底物的泛素化，是控制 SCF 复合体底物选择的决定性因素（Naoki et al., 2004; Kipreos et al., 2000）。

目前，拟南芥已发现 1 000 个具有 F-box 结构的蛋白（Zhang et al., 2002），拟南芥蛋白组中有 700 多个 F-box 蛋白（Gagne et al., 2002; Lechner et al., 2006）。F-box 蛋白在许多信号传导途径中发挥作用（Craig et al., 1999），如生长素、赤霉素、茉莉酸甲酯等（Yu et al., 2007）。拟南芥 F-box 蛋白根据 F-box 结构域的保守性进行家族分类，属于同一亚家族的基因由于 C 端结合底物的特异性在功能上存在很大差异，如 *TIR1*、*COI1* 和 *ORE9*/*MAX2* 属于 C3 亚家族，尽管 F-box 结构域 C 端 LRRs、内含子插入位置相似，但遗传分析其编码的蛋白功能差异很大。拟南芥 *TIR1* 参与生长素信号转导（Gray et al., 1999）；*COI1* 参与茉莉酸甲酯信号转导（Xie et al., 1998）；*ORE9*、*MAX2* 参与叶片衰老和侧枝形成（Woo et al., 2001; Stirnberg et al., 2002）。因此，直接研究同一亚家族成员很有必要。结合我们以前的报道结果：为了探索 ABA 提高辣椒耐寒性的分子机制，我们构建正反交 SSH 文库，并从反交文库中分离了与 F-box 蛋白相似的 EST（Guo et al., 2013）；而且 ABA 预处理的遭受低温胁迫的辣椒叶片 F-box 蛋白基因下调表达，而单一的 ABA 处理和低温胁迫 F-box 蛋白基因上调表达。本试验将进一步研究辣椒 F-box 蛋白基因在非生物胁迫中所起的生物学功能。具体内容为①F-box 蛋白基因在辣椒不同组织中的表达模式分析；②F-box 蛋白基因在激素和非生物逆境胁迫下的表达模式分析，如 SA、高盐、甘露醇和汞胁迫；③发生 F-box 沉默的植株遭受非生物胁迫的抗性分析。

第3章 辣椒对低温胁迫的响应与其低温抗性相关基因的克隆和功能分析

6.1 材料与处理

6.1.1 主要材料及处理方法

本研究拟采用的辣椒材料为 P70 栽培种,由西北农林科技大学辣椒课题组提供。辣椒生长到 6～7 片真叶时,进行 ABA、SA 及汞、低温、盐和干旱处理,处理方法同第 3 章第 4 节描述。

6.1.2 CaF-box 分离和序列分析

以 SSH 反交库 EST(登录号 JZ198812)为探针,按照第 4 章描述的同源克隆法获得基因全长。*CaF-box* 基因全长引物为:

  F:5′-CTTGGTTGTGATTTTCTTGG-3′

  R:5′-CACTCGTGTTT GCTTCTGTA-3′

采用 SMART 软件 (http://smart.embl-heidelberg.de) 分析 F-box 结构域和 LRR 简单重复序列 (Letunic et al., 2009; Schultz et al., 1998)。

6.1.3 病毒沉默 CaF-box 载体的构建

为了保证基因沉默的特异性,我们在辣椒 *CaF-box*(GenBank 登录号:JX402927)的特异区域设计引物,片段长度为 330 bp。将相应的特异片段插入病毒沉默载体 pTRV2,构建流程同第 3 章第 4 节 4.1.3.1 节。所用引物为:

  VCaF-box-F:5′-CTTGGTTGTGATTTTCTTGG-3′

  VCaF-box-R:5′-CGCGGATCCCGCTGCACCCCTTTCAC-3′

(下游引物添加 *Bam* HI 酶切位点和保护碱基,下划线为 *Bam* HI 酶切位点)。

所用限制性内切酶为:*Xba* I 和 *Bam* HI。

6.1.4 辣椒 CaF-box 在非生物胁迫下的表达模式分析

植株生长到 6～7 片真叶时,进行 ABA、SA、低温、盐、渗透和汞处理。RNA 提取、单链 cDNA 合成、实时定量均参考第 3 章第 3 节。

6.1.5 VIGS 技术分析辣椒 CaF-box 沉默的功能

(1) 病毒沉默载体 pTRV2-*CaF-box* 浸染辣椒 参考第 3 章第

4 节 4.1.7 节。子叶完全展开时,用叶片注射法进行感染植株。

(2) 辣椒 $CaF\text{-}box$ 沉默效率的分析方法参考第 3 章第 4 节 4.1.7.1 节。

(3) 盐和渗透胁迫对 $CaF\text{-}box$ 沉默的植株表型影响 取 $CaF\text{-}box$ 沉默的离体叶片（1 cm²），分别放在 300 mmol/L NaCl 和 300 mmol/L 甘露醇溶液中放置 3 d，培养条件同第 3 章第 4 节，观察离体叶片的表型变化。

(4) 低温对 $CaF\text{-}box$ 沉默的植株丙二醛和过氧化氢含量的影响 对 $CaF\text{-}box$ 沉默的植株分别进行叶面喷施 0.57 mmol/L ABA 和清水，72 h 后放置低温 6℃处理 6 h，取样进行丙二醛和过氧化氢含量测定，方法参考第 3 章第 2 节。

6.2 结果与分析

6.2.1 辣椒 CaF-box 全长 cDNA 的分离

以 SSH 反交库中筛选得到的 $F\text{-}box$ 同源 EST（R028）为探针，参考第 3 章第 4 节 4.1.2 节描述的同源克隆法获得基因全长，命名为 $CaF\text{-}box$。该基因的 cDNA 全长为 2 088 bp（GenBank 登录号：JX402925），开放阅读框为 1 914 bp，编码 638 个氨基酸，分子量为 67.8 KD，等电点为 6.96。基因测序获得的 cDNA 全长及其氨基酸序列如图 3-49。

辣椒 $CaF\text{-}box$ 编码的蛋白质结构分析如图 3-50 所示，该基因含有典型的 F-box 蛋白结构域：N 端 63～104 aa 处有 F-box 保守结构域，C 端有 15 个串联 LRR motifs；SlEBF1、SlEBF1、AtEBF2 和 AtEBF2 也含有相应的结构域。而且，$CaF\text{-}box$ 编码的氨基酸与 SlEBF2、PtF-box、SlEBF1 和 AtEBF2 同源率分别为 85.2%、66.0%、55.2%和 50.3%。

进化树分析辣椒 CaF-box 与其他植物如拟南芥、番茄、水稻和毛果杨的 EBF、TIR1、VFB 和 COI1 氨基酸，比对结果表明（图 3-51），CaF-box 属于 F-box 超家族的 EBF 分枝。

第3章 辣椒对低温胁迫的响应与其低温抗性相关基因的克隆和功能分析

```
1 CTTGGTTGTGATTTTCTTGGTTCAAAGTTTGATCTTTCTTGCTTTACATTTTGAGTTTTCTCGACAAGAAGAACAGTC
79 ATGCCTGCTCTTGTTAATTACAGTGGTGATGATGAGTTGTATTCTGGCGGATCATTTTGTCTGCAGATTTGGGTCTCATTGTCTTTGGGT
    M  P  A  L  V  N  Y  S  G  D  D  E  L  Y  S  G  G  S  F  C  S  A  D  L  G  L  M  L  S  L  G
172 CATGCTGAAGTTTACTGTCCTCCTAGGAAGAGGTCGCGCATTTCTGGCCCGTTTGTTGTTGAAGACCGATCAAAGGGTCCATCTCTCGAT
    H  A  E  V  Y  C  P  P  R  K  R  S  R  I  S  G  P  F  V  V  E  D  R  S  K  G  P  S  L  D
262 GATCTTCCCGATGAATGCCTCTTCGAGATTCGACGATTGCCAGGAGGCCGTGAACGAGGTGCAGCCTGTTTGTCTAAAAGATGG
    D  L  P  D  E  C  L  F  E  I  R  R  L  P  G  G  R  E  R  G  A  A  S  C  L  S  K  R  W
352 CTTATGCTGTTGAGTAGCGTCCGGAGCTCTGAAATTTGTAGGAGCAAGAGCTATACCAATCTAAATGACTCAACATGATCTCTAAAGAT
    L  M  L  L  S  S  V  R  S  S  E  I  C  R  S  K  S  Y  T  N  L  N  D  S  T  M  I  S  K  D
442 GAAGATCTTGAAGTTGAATGCGATGGATACCTTACTAGGTGTGTGGAAGGGAAGAAAGCTACTGATGTTAGACTTGCTGCTATTGCAGTT
    E  D  L  E  V  E  C  D  G  Y  L  T  R  C  V  E  G  K  K  A  T  D  V  R  L  A  A  I  A  V
532 GGGACTTCTACCCGCGGAGGACTAGGAAAGCTTTCTGTCCGAGGGAGTAACTCAGTTCGTGGTATCACTAATGTTGGTCTATCCGCGATC
    G  T  S  T  R  G  G  L  G  K  L  S  V  R  G  S  N  S  V  R  G  I  T  N  V  G  L  S  A  I
622 GCGCATGGTTGCCCTTCTCTCAGGGCCCTATCATTGTGGAATGTTCCTTGTATTGGAGATGAAGGTCTTGGAAGTTGCAAGGGAATGC
    A  H  G  C  P  S  L  R  A  L  S  L  W  N  V  P  C  I  G  D  E  G  L  L  E  V  A  R  E  C
712 CGTTCATTAGAAAAGCTAGATCTAAGCCATTGCCCTTCAATCTTCAAGCTGGGTCTTGTCGAGTCAGCCAAGATTGCCCGAGCTTGACT
    R  S  L  E  K  L  D  L  S  H  C  P  S  I  S  N  R  G  L  V  A  I  A  E  N  C  P  S  L  T
802 TCGTTGACAATTGAATCTTGTCCCAATATTGGGAACGAGGGCTTGCAAGCTATTGGAAGATGTTGCAACAACTACAGTTCTTACAATT
    S  L  T  I  E  S  C  P  N  I  G  N  E  G  L  Q  A  I  G  R  C  C  N  K  L  Q  S  L  T  I
892 AAGGACTGTCCTGTTGGGGATCAGGGAATTGCAATCTTGTCATCGGGATCCATGCTAACAAAAGTGGAACTACACTGTCTA
    K  D  C  P  L  V  G  D  Q  G  I  A  S  I  L  S  S  G  A  S  M  L  T  K  V  E  L  H  C  L
982 AACATCACAGATTTCTCCCTTGCAGTCATTGGTCACTATGCAAGCAGATTACAATCTGTGTCTTAGTTCACTTCGTAATGTGAGTCAG
    N  I  T  D  F  S  L  A  V  I  G  H  Y  G  K  Q  I  T  N  L  C  L  S  S  L  R  N  V  S  Q
1072 AAGGGATTTTGGGTCATGGGTAATGCCAAGGGTCTCCAGTCTCTGGTTTTCTTGCAATCACCCTATGCTGGGGAGCCACAGATGTCGGT
    K  G  F  W  V  M  G  N  A  K  G  L  Q  S  L  V  S  L  T  I  T  L  C  W  G  A  T  D  V  G
1162 CTTGAAGCAGTTGGAAAGGGTTGCCCAAATCTTAAACGTATGCATTGCATCAGGTGTCATTGTTTCCGACTGCGGAGTTGTTGCTTTC
    L  E  A  V  G  K  G  C  P  N  L  K  R  M  C  I  R  K  C  C  I  V  S  D  C  G  V  V  A  F
1252 GCTAAAGAGGCTGGATCTCTTGAGTGCTTAAACTTGGAGGAGTGCAACAGGATTACCCAAATAGGTATCTTAACGCAGTTTCAAACTGC
    A  K  E  A  G  S  L  E  C  L  N  L  E  E  C  N  R  I  T  Q  I  G  I  L  N  A  V  S  N  C
1342 AGGAGGTTAAAGAGTCTTTCCCTAGTGAAGTGCATGGGAATCAAAGATCTAGCTCAAACTTCCTTGTTATATCCATGTGAGTCTTTG
    R  R  L  K  S  L  S  L  V  K  C  M  G  I  K  D  L  A  L  Q  T  S  L  L  Y  P  C  E  S  L
1432 CGGTCCTTGTCTATCCGAAGCTGTCCAGGGTTCGGGAGTACTAGCTTAGCTATGATTGGTAAGCTCTGCCCTAAGCTGCATAAATTAGAT
    R  S  L  S  I  R  S  C  P  G  F  G  S  T  S  L  A  M  I  G  K  L  C  P  K  L  H  K  L  D
1522 CTCAGCGGGCTCTGTGGAATAACCGATGCTGGTCTTTCTCCCACTCTTGAGAGCTGTGAAGGACTTGTTAAGGTGAATCTTAGTGACTGC
    L  S  G  L  C  G  I  T  D  A  G  L  L  P  L  L  E  S  C  E  G  L  V  K  V  N  L  S  D  C
1612 CTGAACTTGACAGATCAAGTGGTGCTCTCGTTGGCTGCACGACATGGTGAAGACCCTTGAATTGCTGAATCTTGATGGATGCAGGAAGGTT
    L  N  L  T  D  Q  V  V  L  S  L  A  A  R  H  G  E  T  L  E  L  L  N  L  D  G  C  R  K  V
1702 ACGGATGCAAGTTTGGTGGCAATTGCAGATAATTGCTCATTACTTAATGATCTTGATGTTTCTAAGTGTGCAATACTGATTCTGGTGTA
    T  D  A  S  L  V  A  I  A  D  N  C  S  L  L  N  D  L  D  V  S  K  C  A  I  T  D  S  G  V
1792 GCAGCTTTATCCGTGGAGTGCAAGTGAATTTGCAAGGTGCTCGTTTGGCTATGCCAAATAGTTCTCCTCTT
    A  A  L  S  R  G  V  Q  V  N  L  Q  V  L  S  L  S  G  C  S  M  V  S  N  K  S  V  P  S  L
1882 AAAAAGTTGGGAGAGTGTCTACTTGGTCTGAATCTCCAACATTGCTCCATTAGTTGCAGTTCAGTCGAGCTCTCCGCGGAGGACTTGTGG
    K  K  L  G  E  C  L  L  G  L  N  L  Q  H  C  S  I  S  C  S  S  V  E  L  L  A  E  D  L  W
```

图 3-49 辣椒 *CaF-box* cDNA 及其推导的氨基酸序列

6.2.2 辣椒 *CaF-box* 的表达模式分析

（1）*CaF-box* 在辣椒不同组织的表达分析 分析一些特殊基因的组织表达特异性或许可以为预测该基因的生理功能提供重要线索。qRT-PCR 结果表明 *CaF-box* 在辣椒不同组织中均有表达

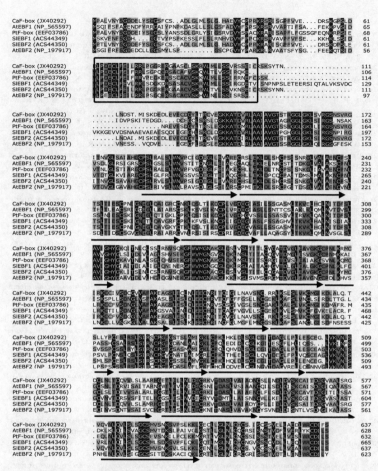

图 3-50 辣椒 CaF-box 与其他 F-box 蛋白的氨基酸序列比对
拟南芥（AtEBF1 和 AtEBF2），番茄（SlEBF1 和 SlEBF2）和毛果杨（PtF-box）。方框里表示结构域，箭头处表示 15 个串联重叠 LRR 序列

（图 3-52），在茎和幼嫩的种子中表达量较高，在根中很低。

（2）CaF-box 响应逆境胁迫下的表达分析 植物通过引发各种生理、生化和分子方面的变化响应环境条件，最终导致基因表达改变。因为非生物逆境影响细胞基因的机制，蛋白降解途径中的组分可能也受到影响，如 F-box 蛋白。为了解决这一问题，对遭受逆境

第3章 辣椒对低温胁迫的响应与其低温抗性相关基因的克隆和功能分析

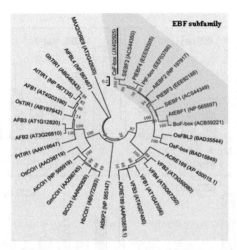

图 3-51 辣椒 CaF-box 与其他物种的 F-box 蛋白的进化树分析

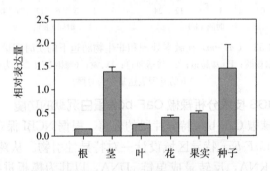

图 3-52 辣椒 *CaF-box* 的组织特异性表达

胁迫下的辣椒幼苗 F-box 蛋白基因的表达模式进行 qRT-PCR 分析，如高盐、甘露醇和重金属汞及植物激素 SA。

SA 诱导 *CaF-box* 表达，1 h 增加到峰值（高于对照 4 倍）（图 3-53A）。NaCl 模拟的高盐胁迫 *CaF-box* 表达变化较大：1 h 增加到对照的 2.5 倍，之后低于对照（图 3-53B）；在甘露醇模拟的高渗胁迫和重金属汞处理下 *CaF-box* 表达均受到抑制（图 3-53C、D），渗透胁迫 24 h 其表达量比对照低 68%。

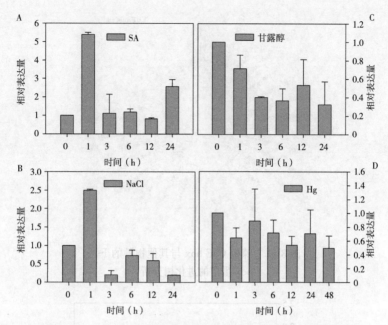

图 3-53 *CaF-box* 在激素处理和非生物胁迫下的实时定量分析
A. 水杨酸；B. 盐胁迫；C. 渗透胁迫；D. 汞；（辣椒 *UBI-3* 作为内参；
0 h 的基因表达量作为对照）

6.2.3 VIGS 技术分析辣椒 CaF-box 蛋白沉默的功能

（1）辣椒 CaF-box 特异片段的分离 根据 NCBI 保守域比对的结果，在 *CaF-box* 特异区域设计一对特异性引物。从辣椒 P70 叶片提取总 RNA，反转录成单链 cDNA，以其为模板进行 PCR 扩增，扩增产物大小为 330 bp。将其连入 pMD19-T 载体进行测序，测序结果与全长基因的特异区域完全一致。

（2）病毒沉默载体的构建 取测序正确的 pMD19-T-*CaF-box* 质粒和空载体 pTRV2：00，分别进行 *Xba* I 和 *Bam* HI 双酶切。回收相应大小的片段，进行 16℃过夜连接。酶切鉴定大肠杆菌转化子质粒，片段大小与预期长度一致，说明病毒载体 pTRV2-*CaF-box* 已成功构建（图 3-54）。然后，将转化子质粒转入农杆菌感受态 GV3101。菌液 PCR 检测转入 pTRV2-*CaF-box* 的菌株，以空载体 pTRV2：00 质粒作为阴性对照，结果证明以农杆菌为模板，PCR 扩增出与预期目的片段大小一致的片段，以空载体质粒为模板没有出现条带（图 3-55）。

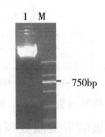

图 3-54 病毒载体 pTRV2-*CaF-box* 双酶切检测

泳道 1：pTRV2-*CaF-box* 双酶切产物；M：2 Kb 分子标记

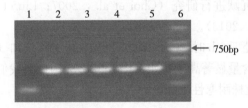

图 3-55 病毒沉默载体 pTRV2-*CaF-box* 的菌液 PCR 检测

泳道 1：以空载体质粒 pTRV2 为模板的 PCR 产物；泳道 2-6：转化 pTRV2-*CaF-box* 质粒的菌液 PCR 产物；M：2Kb 分子标记

（3）*CaF-box* 沉默效率分析

①pTRV2-*PDS* 沉默结果与检测。辣椒 *PDS* 沉默检测方法及其表型分析见第 3 章第 4 节 4.2.6.3 节。

②pTRV2-*CaF-box* 沉默效率分析。在病毒诱导基因沉默系统中，插入的外源基因会随病毒载体在植物各组织中大量转录，因此，我们在病毒载体包含的目的基因的外围区域合成一对特异引物（F：5′- ATACCAATCTAAATGACTCAACAA-3′；R：5′- CCAACATTAGTGATACCACGAAC-3′）。农杆菌接种 30 d 后，调查单株第 5、6 叶位中的 *CaF-box* 转录水平。与阴性对照叶片相比，pTRV2-*CaF-box* 感染植株 *CaF-box* 的表达量下降 60.3%（图 3-56）。

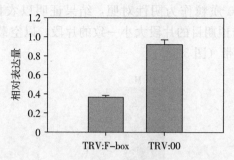

图 3-56 qRT-PCR 分析 *CaF-box* 的沉默效率

③*CaF-box* 沉默对辣椒响应低温处理的抗性分析。经 qRT-PCR 分析，在低温胁迫下辣椒 *CaF-box* 表达上调。为了进一步分析 *CaF-box* 对植株耐寒性的影响，我们采用病毒诱导沉默技术诱导基因发生沉默进行研究（Choi et al.，2007；Liu et al.，2002；Wang et al.，2013）。

在低温处理 6 h 时，*CaF-box* 沉默植株电导率增加（图 3-57A），而且丙二醛含量显著高于对照（图 3-57B）。因此，我们推测 *CaF-box* 沉默使植株耐寒性下降。

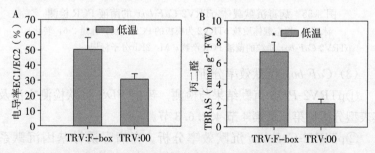

图 3-57 低温对 *CaF-box* 沉默辣椒电导率和丙二醛含量的影响

④*CaF-box* 沉默对辣椒响应盐胁迫和渗透胁迫的抗性分析。经 qRT-PCR 分析，在高盐和渗透胁迫下辣椒 *CaF-box* 表达下降。为了进一步分析 *CaF-box* 沉默对植株耐逆性的影响，我们也采用病毒诱导沉默技术进行研究（Choi et al.，2007；Liu et al.，2002；Wang et al.，2013）。

靠近发生基因沉默叶片的上部叶片（第5或6叶位）进行氯化钠和甘露醇胁迫处理。离体圆形叶片漂浮在无菌水和300 mmol/L NaCl模拟的盐胁迫3 d，无菌水处理 *CaF-box* 沉默离体叶片变黄，而300 mmol/L NaCl处理表现明显失绿，特别是叶脉处和离体叶片边缘，而注射空载体的阴性对照植株叶片出现轻微的光漂白现象（图3-58）。300 mmol/L甘露醇处理 *CaF-box* 沉默植株与清水处理相比，出现更明显的黄化现象。因此，我们推测高盐和渗透胁迫下辣椒 *CaF-box* 的沉默促进植株叶片黄化，进而降低抗性。

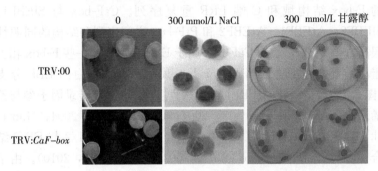

图3-58 *CaF-box* 沉默的辣椒在盐和渗透胁迫下的表型变化

6.3 讨论

F-box蛋白参与激素信号转导的机制研究较为深入的是乙烯、生长素、赤霉素和茉莉酸甲酯（Yu et al.，2007）。拟南芥F-box蛋白（COI1）参与JA调节的防御反应。RPD3b是JA信号转导的负调控因子，COI1通过结合并降解RPD3b进行调节，参与JA信号转导的基因表达（Devoto et al.，2002）。很多JA和机械损伤诱导的基因表达需要COI1（Devoto et al.，2005）。F-box蛋白基因（*SON1*）通过泛素蛋白途径调控防御反应（Kim and Delaney，2002）。然而F-box蛋白基因与ABA信号转导的关系研究较少。在干旱胁迫下，拟南芥 *DOR* 负调控ABA诱导的气孔关闭（Zhang et al.，2008），*DOR* 编码F-box蛋白，属于AtSFL（S-locus-F-box-like）家族成员。而且关于F-box蛋白的研究主要集中在模式植物拟南芥上，在其他作物上报道很少（Yu et al., 2007）。本试验以经济作物辣椒为材料，主要研究辣椒叶片 CaF-box 的表达调控与抗逆性的关系，以及是否参与 ABA 信号转导。

拟南芥中具有 LRR 蛋白互作结构的 F-box 蛋白组成一个大的 C 家族（Gagne et al., 2002），该家族中的很多蛋白调节激素反应：如 TIR1、COI1、EBF1 和 EBF2（Bai et al., 1996；Dharmasiri et al., 2005；Feys et al., 1994；Guo and Ecker, 2003；Kepinski and Leyser, 2005）。辣椒 CaF-box 编码的氨基酸分别在 N 端含有典型的 F-box 结构域和 C 端 LRR 重复序列。CaF-box 与 SlEBF1、SlEBF2、AtEBF1、AtEBF2 和 PtF-box 之间序列和结构的同源性很高（图 3-44），表明 CaF-box 属于 EBF 亚家族。一些 F-box 相关蛋白的聚类分析也证明了 CaF-box 属于 F-box 蛋白的 EBF 分支（图 3-45）。拟南芥 AtEBF1/2 通过直接降解 EIN3 转录因子参与乙烯信号转导（Binder et al., 2007；Gagne et al., 2004；Guo et al., 2003；Potuschak et al., 2003）。番茄 Sl-EBF1/2 与 EIN3 结合，参与乙烯反应（Guo et al., 2003；Yang et al., 2010）。由于 CaF-box 与 SlEBF1、AtEBF1、AtEBF2 的 C 端序列和结构高度同源，因此，我们推测辣椒 CaF-box 也与乙烯信号转导有关。

在植物的不同组织中，植物 EBF 类基因的表达模式各不相同，如番茄 Sl-EBF1/2 在花中表达量最高（Yang et al., 2010），而拟南芥 AtEBF1 和 AtEBF2 在各种组织中组成型表达（Potuschak et al., 2003）。本试验 CaF-box 在辣椒根、茎、幼叶、花、青熟期果实和幼嫩种子等组织均有表达，茎和种子表达量较高（图 3-46）。植物逆境反应通常模拟某些正常的发育过程（Cooper et al., 2003）。植物发育和环境条件之间的关系表明有些基因受环境因子和生长因素的共同调节。番茄磷脂酶 D 和产物磷脂酸在花粉萌发和花粉管伸长中发挥作用（Potocky et al., 2003；Zonia et al., 2004），而且也参与各种逆境反应如失水、高盐和机械损伤（Wang, 2002）。在水稻上逆境反应和种子发育之间存在交叉基因（Cooper et al., 2003）。

研究报道 ABA 诱导葡萄 F-box 蛋白（*BIG-24.1*）表达（Paquis et al.，2011）；菜豆 F-box 蛋白（*PvFBS*1）被 ABA 和 SA 处理上调表达（Maldonado-Calderón et al.，2012）。本试验中 ABA 和 SA 处理也诱导辣椒 *CaF-box* 表达（图 3-47A）。尽管 SA 通常被认为参与生物胁迫调节，而 ABA 在非生物胁迫中起着重要的作用，但两者之间存在着复杂的正协调和负协调关系（Robert-Seilaniantz et al.，2011）。Jain et al.（2007）研究表明遭受非生物逆境胁迫的水稻 43 个 F-box 蛋白存在差异表达，如低温、高盐和失水胁迫。7 个 F-box 蛋白基因在以上提到的逆境胁迫上调表达，12 个在任何两种胁迫下存在差异表达，如在高盐和失水胁迫下 9 个和 2 个分别上调和下调表达。另外，低温诱导辣椒 *CaF-box* 表达（图 3-48C），与低温对拟南芥 *AtFBS*1 表达的影响一致（Matsui et al.，2008）。高盐处理 1 h *CaF-box* 表达快速增加（图 3-47B）；而渗透胁迫和汞处理辣椒 *CaF-box* 下调表达（图 3-47C 和 D）；以上结果表明在不同的逆境条件下，该基因的调控表现不同。

为了进一步分析辣椒 *CaF-box* 在非生物胁迫中所起的作用，对 *CaF-box* 沉默植株进行低温、高盐和渗透胁迫处理。与未发生基因沉默的植株相比，低温 6 h *CaF-box* 沉默的植株丙二醛和电导率显著增加（图 3-51），表明 *CaF-box* 沉默的辣椒耐寒性减弱。据报道乙烯在拟南芥响应低温胁迫的反应中扮演负调控因子的角色（Shi et al.，2012）。我们推测 *CaF-box* 沉默后，辣椒在低温胁迫下产生乙烯反应，耐寒性减弱。乙烯信号转导参与高盐和渗透胁迫反应。EIN3 在高盐胁迫中起着正调节作用，如在高盐胁迫下，双突变体 *EBF1/2* 幼苗 EIN3 大量积累，耐盐性增强；过量表达 *EIN3* 株系具有乙烯组成型表型，耐盐性也增强（Achard et al.，2006）。高盐和渗透胁迫抑制拟南芥乙烯受体 *ETRI* 表达，从而激活乙烯反应基因的表达（Zhao et al.，2004）。本试验在高盐和渗透胁迫下，*CaF-box* 沉默离体叶片出现明显的黄化现象（图 3-52）。叶绿素下降是衰老的重要指标，因此我们推测，*CaF-box* 沉默植株在高盐和渗透胁迫下表现早衰现象。同一 EBF 分支上的高同源 *SlEBF1* 和 *SlEBF2* 沉默后，植株表现典型的乙烯反应，即促进衰老（Yang et al.，2010）。因此我们推测辣椒 *CaF-box* 在高盐和渗透胁迫下参与乙烯信号转导。

在植物一些组织和发育过程中，ABA 与乙烯信号途径之间多表现为拮抗作用，如在种子发芽和根系生长期间乙烯信号转导负调节 ABA 信号（Ghassemian et al.，2000）；乙烯通过降低对 ABA 的敏感性促进种子发芽，另外 ABA 抑制根系生长需要乙烯参与（Beaudoin et al.，2000）；乙烯抑制 ABA 诱导的气孔关闭（Tanaka et al.，2005）。然而高浓度的生长素激发乙烯诱导 ABA 积累，然后 ABA 作为第二信使，产生一些形态变化，如抑制生长，促进衰老（Hansen et al.，2000），然而逆境条件下乙烯的积累诱导 ABA 合成增加，从而介导逆境条件下出现的衰老现象值得研究。ABA 与乙烯存在复杂的拮抗和协同关系，本试验研究表明ABA 调节辣椒 *CaF-box* 表达，但乙烯是否参与调节该基因表达，以及这两种激素之间的关系有待于进一步研究。

7 全文结论及创新点

7.1 全文结论

（1）低温启动辣椒 AsA-GSH 循环代谢进行活性氧的清除。外源 ABA 可提高辣椒幼苗 SOD 和 POD 酶活及其基因表达，进而增强对低温诱导氧化胁迫的抵抗性。

（2）对 ABA 预处理的遭受低温胁迫的辣椒幼苗进行抑制消减杂交文库（SSH）构建。共获得 235 个高质量的 ESTs，为筛选低温抗性相关基因提供了资源。

（3）基于 SSH 文库分离的差异基因，通过同源克隆法获得 5 个与低温抗性相关的 cDNA 全长，分别命名为 *CaNAC2*、*CaMBF*、*CaF-box*、*CaMADS-box*，GenBank 登录号分别为 JX402928、JX402927、JX402925、JX402926。

第3章　辣椒对低温胁迫的响应与其低温抗性相关基因的克隆和功能分析

（4）辣椒 $CaNAC2$ 功能缺失加重了低温引起的膜脂过氧化的程度，而在高盐胁迫下，$CaNAC2$ 沉默的离体叶片保持较绿。

（5）辣椒 $CaMBF$ 转基因拟南芥的耐寒性减弱，与 $CaMBF$ 负调节一些逆境防御基因有关，如 $RD22$、$ERD15$、$RD29A$ 等。

（6）辣椒 CaF-box 功能缺失引起植株的耐寒性减弱。在高盐和渗透胁迫下，CaF-box 沉默的离体叶片出现明显的黄化现象。

7.2　创新点

（1）从辣椒上分离235个高质量的低温抗性相关ESTs序列，并从中克隆 $CaNAC2$、$CaMBF$、CaF-box、$CaMADS$-box 等4个低温抗性相关的cDNA全长序列，为挖掘低温抗性相关基因奠定了基础。

（2）利用VIGS技术和转基因方法对 $CaNAC2$、$CaMBF$、CaF-box 进行了功能分析，结果表明这3个基因参与辣椒对逆境的抗性反应，如低温、盐胁迫，为进一步明确这些基因的功能提供了科学依据。

（3）明确了外源ABA可有效地提高辣椒的耐寒性，丰富了辣椒遭受低温胁迫后抗坏血酸-谷胱甘肽循环的代谢理论。

参 考 文 献

蔡庆生，2013. 植物生理学实验 [M]. 北京：中国农业大学出版社.
陈娜，郭尚敬，孟庆伟，2005. 膜脂组成与植物抗冷性的关系及其分子生物学研究进展 [J]. 生物技术通讯 (2)：6-9.
陈年来，张玉鑫，王春林，等，2011. BTH 处理对甜瓜叶片活性氧代谢的诱导作用 [J]. 植物保护学报，6：499-505.
陈夕军，朱键鑫，陈羽，等，2015. 抗白粉病黄瓜品种的叶片组织结构及其生理生化 [J]. 江苏农业学报，31 (1)：55-61.
程鸿，2009. 甜瓜 *APX* 和 *Mlo* 基因的克隆与功能分析 [D]. 泰安：山东农业大学.
崔慧萍，周薇，郭长虹，2017. 植物过氧化物酶体在活性氧信号网络中的作用 [J]. 中国生物化学与分子生物学报，33 (3)：220-226.
邓丽君，廖新福，王惠林，等，2015. 甜瓜白粉病离体叶片接种方法研究 [J]. 中国瓜菜，28 (3)：18-21.
刁倩楠，曹燕燕，蒋雪君，等，2021. 白粉病菌对不同甜瓜品种幼苗生理生化指标的影响 [J]. 分子植物育种，19 (7)：2346-2353.
段学武，蒋跃明，2005. 高氧对果蔬采后生理影响研究进展 [J]. 热带亚热带植物学报，13 (6)：543-548.
付健，王玉凤，张翼飞，等，2021. 不同木霉菌对寒地盐碱土壤玉米杂交种光合特性及活性氧代谢的影响 [J]. 黑龙江八一农垦大学学报，33 (1)：7-14.
傅仕敏，2014. 柑橘黄龙病的细胞病理及其寄主转录组学研究 [D]. 重庆：西南大学.
高俊杰，秦爱国，于贤昌，2009. 低温胁迫对嫁接黄瓜叶片抗坏血酸—谷胱甘肽循环的影响 [J]. 园艺学报，36 (2)：215-220.
高丽华，2007. 南瓜 NBS 类抗病基因同源序列的克隆与分析 [D]. 福州：福建农林大学.
高婷婷，2013. 大白菜抗霜霉病防御信号途径中相关基因的表达分析 [D]. 扬州：扬州大学.

参 考 文 献

韩欢欢,马韬,谢冰,2012. 瓜类蔬菜白粉病抗性诱导及其抗病机制研究进展 [J]. 中国农学通报,28 (25):124-128.

韩正敏,尹佟明,1998. 杨树过氧化物酶活性,气孔密度和大小与黑斑病抗性的关系 [J]. 南京林业大学学报,22 (4):91-93.

郝欣,2016. 转 $hpa1_{Xoo-N21}$ 基因烟草三种抗病信号通路相关基因研究 [D]. 扬州:扬州大学.

何金环,李存法,梁月丽,等,2005. 植物细胞活性氧及其胞内信号转导 [J]. 河南农业科学,8:18-20.

金海军,胡继军,卜立君,等,2020. 华南型黄瓜自交系白粉病抗性与光合特性的关系 [J]. 中国瓜菜,33 (2):27-30.

柯思佳,毕研飞,任琴琴,等,2019. 蔓枯病菌侵染影响抗感甜瓜品种及杂种的光合特性分析 [J]. 植物病理学报,49 (4):465-473.

黎豪,龚诗琦,宋慧娟,等,2021. 南瓜白粉病研究进展 [J]. 中国瓜菜,34 (2):8-11.

李冬林,金雅琴,崔梦凡,等,2019. 遮光对连香树幼苗光合特性及其叶片解剖结构的影响 [J]. 西北植物学报,39 (6):1053-1063.

李建武,2010. 黄瓜霜霉病抗性相关基因筛选及过敏性抗病机制 [D]. 武汉:华中农业大学.

李明玉,曹辰兴,于喜艳,2006. 低温锻炼对冷胁迫下黄瓜幼苗保护性酶的影响 [J]. 西北农业学报,15 (1):160-164.

李宁,王美月,孙锦,等,2013. 外源 24-表油菜素内酯对弱光胁迫下番茄幼苗生长及光合作用的影响 [J]. 西北植物学报,33 (7):1395-1402.

李佩芳,2013. 黄瓜霜霉病过敏性反应的研究 [D]. 郑州:河南农业大学.

李响,戴懿,梁鹏,等,2013. 橡胶树白粉菌诱导寄主及非寄主植物产生活性氧积累的研究 [J]. 广东农业科学,40 (9):68-70.

李小玲,刘长命,刘炼红,等,2015. 外源亚精胺对甜瓜幼苗白粉病抗性的影响 [J]. 西北植物学报,35 (9):1800-1807.

李新峥,杜晓华,张振伟,2009. 中国南瓜经济性状遗传初探 [J]. 西北农业学报,18 (4):319-323.

李征,刘登义,王育鹏,等,2006. 活性氧在植物—病原物相互作用过程中的作用 [J]. 安徽师范大学学报:自然科学版,29 (1):70-74.

梁巧兰,徐秉良,颜惠霞,等,2010. 南瓜白粉病病原菌鉴定及寄主范围测定 [J]. 菌物学报,29 (5):636-643.

林金明,屈锋,单孝全,2002. 活性氧测定的基本原理与方法 [J]. 分析化学,30 (12):1507-1514.

刘鸿先，曾韶西，王以柔，等，1985. 低温对不同耐冷力的黄瓜幼苗子叶各细胞器中超氧物歧化酶的影响 [J]. 植物生理与分子生物学学报，11：48-57.

刘会宁，朱建强，万幼新，2001. 几个欧亚种葡萄材料对霜霉病的抗性鉴定 [J]. 上海农业学报，17 (3)：64-67.

刘龙洲，2008. 黄瓜白粉病抗性遗传分析和基因定位研究 [D]. 上海：上海交通大学农业与生物学院.

刘鹏，李勃，刘庆忠，等，2003. 冷锻炼诱导甜椒抗冷力的生化机理研究 [J]. 山东农业科学，3：11-14.

刘淑艳，王丽兰，姜文涛，等，2011. 中国长春瓜类白粉菌 *Podosphaera xanthii* 形态学和分子系统学研究 [J]. 菌物学报，30 (5)：702-712.

刘文君，王日升，黄凤婵，等，2013. 密本南瓜白粉病抗性遗传规律分析 [J]. 北方园艺，(14)：39-41.

刘新社，陈妍，2019. 生物农药对设施和露地黄瓜白粉病的防效及其品质和产量的影响 [J]. 河南农业科学，48 (3)：95-99.

刘阳，2007. 番茄 MBF1 转录辅激活因子基因的克隆载体的构建及转化番茄 [D]. 重庆市：重庆大学.

刘洋，屈淑平，崔崇士，2006. 南瓜营养品质与功能成分研究现状与展望 [J]. 中国瓜菜，(2)：27-29.

柳利龙，梁巧兰，沈慧敏，2014. 4 种化学物质对南瓜抗白粉菌产生过敏性反应影响的研究 [J]. 植物保护，40 (4)：26-31.

芦光新，胡东维，徐颖，2016. 活性氧积累与小麦抗白粉病关系的研究 [J]. 麦类作物学报，26 (4)：149-153.

马艳青，戴雄泽，2000. 低温胁迫对辣椒抗寒性相关生理指标的影响 [J]. 湖南农业大学学报（自然科学版），26 (6)：461-462.

孟大山，2005. 如何防治南瓜白粉病 [J]. 农业知识，(20)：5.

潘汝谦，2004. 黄瓜对霜霉病的抗性与叶绿素含量、气孔密度的相关性 [J]. 上海交通大学学报，22 (4)：381-384.

裴冬丽，马原松，李成伟，2010. 番茄与白粉病菌互作反应中活性氧的积累 [J]. 植物保护学报，37 (1)：91-92.

齐晓花，罗晶晶，Mouammar Alfandi，等，2010. 黄瓜抗白粉病品系 cDNA 文库构建及 EST 分析 [J]. 园艺学报，37 (6)：931-938.

祁艳，王冬梅，2007. 植物中活性氧的检测方法 [J]. 生物技术通报，5：72-75.

乔禹，丁国华，2016. NPR 在 SA 信号通路中的研究进展 [J]. 中国农学通报，32 (8)：37-43.

参考文献

秦宏坤,李帅,马冬,等,2020. 南瓜侵染白粉菌后的活性氧暴发观察 [J]. 中国农学通报,36(25):121-128.

屈淑平,綦聪,王云莉,等,2018. 南瓜白粉病病原菌及种质资源抗性鉴定 [J]. 东北农业大学学报,49(12):9-17.

任旭琴,张林青,孙敏,2006. 辣椒叶片对低温的生理响应研究 [J]. 安徽农业科学,34(24):6439-6440.

沈喜,李红玉,2003. 病毒侵染后植物叶绿体光合作用变化的分子机制 [J]. 植物病理学报,33(4):289-291.

司胜伟,2011. 依据过敏性反应划分黄瓜抗霜霉病类型 [D]. 郑州:河南农业大学.

苏维埃,1998. 植物对温度逆境的适应 [M]. 植物生理与分子生物学. 北京:科学出版社,721-727.

汪承润,何梅,李月云,2012. 植物体超氧阴离子自由基不同检测方法的比较 [J]. 环境化学,31(5):726-730.

王达菲,2009. 番茄转录辅激活子 LeMBF1 在转基因材料中的功能分析 [J]. 重庆市:重庆大学.

王建设,陈杭,2000. 甜瓜抗白粉病鉴定 [J]. 华北农学报,15(1):125-128.

王教敏,付卫东,吴云,等,2009. 臭椿炭疽病菌盘长孢刺盘孢 (*Colletotrichum gloeosporioides* Penz.) 生物学特性研究 [J]. 山东农业大学学报:自然科学版,40(1):27-31.

王玲平,2001. 黄瓜感染枯萎病菌后生理生化变化及其与抗病性关系的研究 [D]. 山西:山西农业大学.

王敏,2011. SOD1 模型化合物调控干旱胁迫下水稻细胞的抗氧化活性 [D]. 武汉:华中师范大学.

王萍,王宏太,陈鸿,等,2010. 两种方法鉴定甜瓜白粉病抗性的比较 [J]. 长江蔬菜,8:96-98.

王莹,秦阳阳,曾婷,等,2022. 柑橘黄脉病毒侵染对柠檬光合特性和叶绿体超微结构的影响 [J]. 园艺学报,49(4):861-867.

吴仁锋,杨绍丽,2013. 瓜类白粉病及其防治技术 [J]. 长江蔬菜,(21):51.

夏伟伟,邓竹根,戴祖云,等,2021. 长丰地区南瓜白粉病病原菌生理小种鉴定 [J]. 中国瓜菜,34(11):24-29.

咸丰,2012. 野生甜瓜 [sp. *Agrestis* (Naud.) Greb.] 抗白粉病的遗传机制和激素变化及其 cDNA-AFLP 分析 [D]. 杨凌:西北农林科技大学.

向婧姝，2018. 植物过敏性反应分子机制研究进展 [J]. 江西农业学报，30 (12): 41-46.

谢文华，谢大森，1999. 棱角丝瓜不同材料对霜霉病抗性的相关研究 [J]. 华南农业大学学报，20 (2): 28-31.

徐晓晖，孙骏威，郭泽建，2007. 植物与病原菌互作中活性氧的检测方法 [J]. 中国计量学院学报，18 (1): 49-53.

徐正，2018. 茉莉酸路径与脂转运蛋白 GhLTPd 在棉花 ssn 突变体类病斑形成中的功能研究 [D]. 武汉：华中农业大学.

许珂，王萍，崔晓伟，等，2021. 籽用美洲南瓜叶片对白粉病菌的抗逆生理响应 [J]. 西北植物学报，41 (10): 1673-1680.

薛鑫，张芊，吴金霞，2013. 植物体内活性氧的研究及其在植物抗逆方面的应用 [J]. 生物技术通报，10: 6-11.

闫朝辉，2017. 欧洲葡萄 VvDRL1 基因的功能研究 [D]. 河南科技学院.

颜惠霞，2009. 南瓜白粉病材料抗病性及抗病机理研究 [D]. 兰州：甘肃农业大学.

杨广东，郭庆萍，1998. 低温对青椒幼苗过氧化物酶和超氧物歧化酶活性的影响 [J]. 山西农业科学，26 (4): 44-47.

尹延旭，2009. 辣椒抗寒性鉴定技术及其机理研究 [D]. 杨凌：西北农林科技大学.

张海英，苏芳，郭绍贵，等，2008. 甜瓜白粉病抗性基因 Pm-2F 的遗传特性及与其紧密连锁的特异片段 [J]. 园艺学报，35 (12): 1773-1780.

张焕春，尹国香，刘学卿，等，2014. 保护地南瓜白粉病的发生及综合防治技术研究 [J]. 安徽农业科学，42 (2): 431-432.

张梦如，杨玉梅，成蕴秀，等，2014. 植物活性氧的产生及其作用和危害 [J]. 西北植物学报，34 (9): 1916-1926.

张素勤，顾兴芳，张圣平，2008. 黄瓜对霜霉病和白粉病抗性研究 [J]. 江苏农业科学，3: 136-138.

张兆辉，卢盼玲，陈春宏，等，2021. 西葫芦抗白粉病的生理生化机制 [J]. 分子植物育种，19 (9): 3074-3080.

张志良，瞿伟菁，李小方，2009. 植物生理学实验指导 [M]. 高等教育出版社，58-59.

章珍，刘新红，翟洪翠，等，2011. 小麦 Pm21 基因调控的白粉菌早期侵染抑制和寄主细胞反应 [J]. 作物学报，37 (1): 67-73.

赵海新，黄晓群，陈书强，等，2020. 碱胁迫下不同水稻品种微观结构响应解析 [J]. 东北农业大学学报，51 (11): 11-22.

参考文献

赵晓玉, 薛娴, 卢存福, 等, 2014. 植物中活性氧信号转导及其检测方法研究进展 [J]. 电子显微学报, 2: 188-196.

郑东虎, 葛晓光, 张宪政, 等, 2003. 冷胁迫对番茄膜脂过氧化与抗氧化酶系统的影响 [J]. 北方园艺, 4: 46-47.

周俊国, 李桂荣, 杨鹏鸣, 2006. 南瓜自交系数量性状分析与聚类分析 [J]. 河北农业大学学报, 29 (4): 23-26.

周俊国, 李新铮, 张辉蓉, 等, 2011. 中国南瓜自交系 112-2 对白粉病的抗性鉴定 [J]. 北方园艺, (04): 5-8.

周威, 李彩霞, 王飞, 等, 2007. 白粉病菌入侵对不同抗性南瓜品种的病理和生理影响 [J]. 安徽农业科学, 35 (6): 1711-1713.

朱自果, 2012. 中国野生华东葡萄 cDNA 文库测序及转录因子基因 *ERF* 和 *NAC* 功能分析 [D]. 杨凌: 西北农林科技大学.

邹志荣, 陆帼一, 1994. 低温对辣椒幼苗膜脂过氧化和保护酶系统变化的影响 [J]. 西北农业学报, 3 (3): 51-56.

Achard P, Cheng H, De Grauwe L, et al., 2006. Integration of plant responses to environmentally activated phytohormonal signals [J]. Science, 311: 91-94.

Aebi H, 1984. Catalase in vitro [J]. Methods Enzymol, 105: 121-126.

Ahmed S S, Gong Z H, Ji J J, et al., 2012. Construction of the intermediate vector pVBG2307 by incorporating vital elements of expression vectors pBI121 and pBI221 [J]. Genetics and Molecular Research, 11 (3): 3091-3104.

Ai T N, Naing A H, Yun B W, et al., 2018. Overexpression of *RsMYB1* enhances anthocyanin accumulation and heavy metal stress tolerance in transgenic petunia [J]. Front. Plant Sci., 9: 1388.

Airaki M, Leterrier M, Mateos R M, et al., 2012. Metabolism of reactive oxygen species and reactive nitrogen species in pepper (*Capsicum annuum* L.) plants under low temperature stress [J]. Plant Cell Environ, 35: 281-295.

Alabadi D, Oyama T, Yanovsky M J, 2001. Reciprocal regulation between TOC1 and LHY/CCA1 within the Arabidopsis circadian clock. *Science*, 293: 880.

Al-Attala M N, Wang X J, Abou-Attia M A, et al., 2014. A novel TaMYB4 transcription factor involved in the defence response against *Puccinia striiformis* f. sp. tritici and abiotic stresses [J]. Plant Mol. Biol., 84: 589.

Alexander D, Goodman R M, Cut-Rella M, et al., 1993. Increased tolerance to two oomycete pathogens in transgenic tobacco expressing pathogenesis-related protein 1a. Proc Natl Acad Sc. USA., 90 (15): 7327-7331.

Ali S, Mir Z A, Bhat J A, et al., 2018. Isolation and characterization of systemic acquired resistance marker gene *PR1* and its promoter from *Brassica juncea* [J]. 3 Biotech., 8: 10.

Altschul S F, Madden T L, Schaffer A A, et al., 1997. Gapped BLAST and PSI-BLAST: a new generation of protein database search programs. Nucleic Acids Res, 25, 3389-3402.

Alvarez-Buylla E R, Liljegren S J, Palaz S, et al., 2000. MADS-box gene evolution beyond flowers: expression in pollen, endosperm, guard cells, roots and trichomes [J]. Plant Journal, 24: 457-466.

Ando K, Carr KM, Grumet R, 2012. Transcriptome analyses of early cucumber fruit growth identifies distinct gene modules associated with phases of development [J]. BMC Genom, 13: 518.

Andrade M A, Perez-Iratxeta C, Ponting C P, 2001. Protein repeats: structures, functions and evolution [J]. J Struct Biol, 134: 117-131.

Anisimova O K, Shchennikova A V, Kochieva E Z, et al., 2021. Pathogenesis-related genes of *PR1*, *PR2*, *PR4*, and *PR5* families are involved in the response to Fusarium infection in garlic (*Allium sativum* L.). Int. J. Mol. Sci., 22: 6688.

Apel K, Hirt H, 2004. Reactive oxygen species: metabolism, oxidative stress, and signaling [J]. Annual Review of Plant Biology, 55: 373-399.

Arrigoni O, Dipierro S, Borraccino G, 1981. Ascorbate free radical reductase: a key enzyme of the ascorbic acid system [J]. FEBS Lett, 125: 242-244.

Assmann S M, Shimazaki K L, 1999. The multisensory guard cell, stomatal responses to blue light and abscisic acid [J]. Plant Physiol, 119: 809-816.

Averyanov A, 2009. Oxidative burst and plant disease resistance [J]. Frontiers in Bioscience, 1: 142.

Azevedo C, Betsuyaku S, Peart J, et al., 2006. Role of SGT1 in resistance protein accumulation in plant immunity [J]. EMBO J, 25: 2007-2016.

Bai C, Sen P, Hofmann K, et al., 1996. SKP1 connects cell cycle regulators to the ubiquitin proteolysis machinery through a novel motif, the F-box [J]. Cell, 86: 263-274.

Baisakh N, Subudhi P K, Varadwaj P, 2008. Primary responses to salt stress

参 考 文 献

in a halophyte, smooth cordgrass (*Spartina alterniflora* Loisel.) [J]. Funct Integr Genomics, 8: 287-300.

Balazadeh S, Siddiqui H, Allu A D, et al. , 2010. A gene regulatory network controlled by the NAC transcription factor ANAC092/AtNAC2/ORE1 during salt-promoted senescence [J]. Plant J, 62 (2): 250-264.

Bassanezi R B, Amorim L, Bergamin-Filho A, et al. , 2002. Gas exchange and emission of chlorophyll fluorescence during the monocycle of rust, angular leaf spot and anthracnose on bean leaves as a function of their trophic characteristics [J]. J Phyto pathol 150: 37-47.

Beaudoin N, Serizet C, Gosti F, et al. , 2000. Interactions between abscisic acid and ethylene signaling cascades [J]. Plant Cell, 12: 1103-1115.

Behr M, Humbeck K, Hause G, et al. , 2010. The hemi biotroph *Colletotrichum graminicola* locally induces photosynthetically active green islands but globally accelerates senescence on aging maize leaves [J]. Mol Plant-Microbe Interact, 23: 879-892.

Bellaire B A, Carmody J, Braud J, et al. , 2000. Involvement of abscisic acid-dependent and-independent pathways in the upregulation of antioxidant enzyme activity during NaCl stress in cotton callus tissue [J]. Free Radic Res, 33: 531-545.

Berger S, Sinha A K, Roitsch T, 2007. Plant physiology meets phytopathology: Plant primary metabolism and plant-pathogen interactions [J]. J Exp Bot, 58: 4019-4026.

Berrocal-LoboM, Molina A, Solano R, 2002. Constitutive expression of ethylene-response-factor1 in *Arabidopsis* confers resistance to several necrotrophic fungi [J]. Plant Journal, 29: 23-32.

Bertin P, Bouharmont J, Kinet J M, 1996. Somaclonal variation and improvement in chilling tolerance in rice: changes in chilling-induced electrolyte leakage [J]. Plant Breeding, 115: 268-272.

Bhatnagar-Mathur P, Vadez V, Sharma K, 2008. Transgenic approaches for abiotic stress tolerance in plants: Retrospect and prospects [J]. Plant Cell Reports, 27: 411-424.

Bjellqvist B, Hughes G J, Pasquali C, et al. , 1993. The focusing positions of polypeptides in immobilized pH gradients can be predicted from their amino acid sequences. Electrophoresis, 14, 1023-1031.

Blanca J, Esteras C, Ziarsolo P, et al. , 2012. Transcrip-tome sequencing for

SNP discovery across *Cucumis melo* [J]. BMC Genom, 13: 280.

Bleecker A B, Kende H, 2000. Ethylene: A gaseous signal molecule in plants [J]. Annu Rev Cell Dev Biol, 16: 1-18.

Bolton M D, 2009. Primary metabolism and plant defense—fuel for the fire [J]. Mol Plant Microbe Interact, 22 (5): 487-497

Borden S, Higgins V J, 2002. Hydrogen peroxide plays a critical role in the defence response of tomato to *Cladosporium fulvum* [J]. Physiol Mol Plant Pathol, 61: 227-236.

Braun U, 1987. A monograph of the *Erysiphales* (powdery mildews) [J]. Beiheft zur Nova Hedwigia, 89: 1-700.

Bray E A, 1988. Drought- and ABA-induced changes in polypeptide and mRNA accumulation in tomato leaves [J]. Plant Physiol, 88: 1210-1214.

Breen S, Williams S J, Outram M, et al., 2017. Emerging insights into the functions of pathogenesis-related protein 1 [J]. Trends Plant Sci. 22 (10): 871-879.

Brendel C, Gelman L, Auwerx J, 2002. Multiprotein bridging factor-1 (MBF-1) is a cofactor for nuclear receptors that regulate lipid metabolism [J]. Mol Endocrinol, 16: 1367-1377.

Brown R L, Kazan K, McGrath K C, et al., 2003. A role for the GCC-box in jasmonate-madiated activation of the *PDF1.2* gene of *Arabidopsis* [J]. Plant Physiol, 132: 1020-1032.

Buchanan C D, Lim S, Salzman R A, et al., 2005. Sorghum bicolor's transcriptome response to dehydration, high salinity and ABA [J]. Plant Mol Biol, 58: 699-720.

Bueno P, Piqueras A, Kurepa J, et al., 1998. Expression of antioxidant enzymes in response to abscisic acid and high osmoticum in tobacco BY-2 cell cultures [J]. Plant Sci, 138: 27-34.

Caarl L, Pieterse C M J, Wees S C M V, 2015. How salicylic acid takes transcriptional control over jasmonic acid signaling [J]. Front. Plant Sci, 6: 170.

Casano L M, Martín M, Sabater B, 2001. Hydrogen peroxide mediates the induction of chloroplastic Ndh complex under photooxidative stress in barley [J]. Plant Physiology, 125 (3): 1450-1458.

Catala R, Santos E, Alonso J M, et al., 2003. Mutations in the Ca^{2+}/H^+ transporter CAX1 increase CBF/DREB1 expression and the cold-acclimation response in *Arabidopsis* [J]. Plant Cell, 15: 2940-2951.

Chak R K, Thomas T L, Quatrano R S, et al., 2000. The genes *ABI1* and *ABI2* are involved in abscisic acid- and drought-inducible expression of the Daucus carota L. Dc3 promoter in guard cells of transgenic *Arabidopsis thaliana* (L.) Heynh [J]. Planta, 210 (6): 875-883.

Chao Q, Rothenberg M, Solano R, et al., 1997. Activation of the ethylene gas response pathway in *Arabidopsis* by the nuclear protein ethylene-insensitive3 and related proteins [J]. Cell, 89: 1133-1144.

Chen B H, Guo W L, Yang H L, et al., 2020. Photosynthetic properties and biochemical metabolism of Cucurbita moschata genotypes following infection with powdery mildew [J]. Journal of Plant Pathology, 102 (4): 1021-1027.

Chen T Z, Li W J, Hu X H, et al., 2015. A cotton MYB transcription factor, GbMYB5, is positively involved in plant adaptive response to drought stress [J]. Plant Cell Physiol., 56: 917.

Chen X J, He S T, Jiang L N, 2021. An efficient transient transformation system for gene function studies in pumpkin (*Cucurbita moschata* D.) [J]. Sci. Hortic., 282: 110028.

ChenY L, Lee C Y, Cheng K T, et al., 2014. Quantitative peptidomics study reveals that a wound-induced peptide from PR-1 regulates immune signaling in tomato [J]. Plant Cell, 26: 4135-4148.

Cheol Song G, Sim H J, Kim S G, et al., 2016. Root-mediated signal transmission of systemic acquired resistance against above-ground and below-ground pathogens [J]. Annals of Botany, 118: 821-831.

Cheong Y H, Sung S J, Kim B G, et al., 2010. Constitutive overexpression of the calcium sensor CBL5 confers osmotic or drought stress tolerance in *Arabidopsis* [J]. Mol Cells, 29: 159-165.

Chinnusamy V, Jagendorf A, Zhu J, 2005. Understanding andimproving salt tolerance in plants [J]. Crop Science, 45: 437-448.

Chinnusamy V, Zhu J K, Sunkar R, 2010. Gene regulation during cold stress acclimation in plants [J]. Methods Mol Biol, 639: 39-55.

Chinnusamy V, Zhu J, Zhu J K, 2007. Cold stress regulation of gene expression in plants [J]. Trends in Plant Sci, 12: 444-451.

Choi D S, Hwang I S, Hwang B K, 2012. Requirement of the cytosolic interaction between PATHOGENESIS-RELATED PROTEIN10 and

LEUCINE-RICH REPEAT PROTEIN1 for cell death and defense signaling in pepper [J]. Plant Cell, 24: 1675-90.

Choi H W, Kim Y J, Lee S C, et al., 2007. Hydrogen peroxide generation by the pepper extracellular peroxidase *CaPO2* activates local and systemic cell death and defense response to bacterial pathogens [J]. Plant Physiol, 145: 890-904.

Christianson J A, Wilson I W, Llewellyn D J, et al., 2009. The low-oxygen-induced NAC domain transcription factor *ANAC102* affects viability of *Arabidopsis* seeds following low-oxygen treatment [J]. Plant Physiol, 149 (4): 1724-1738.

Clement M, Lambert A, Herouart D, et al., 2008. Identification of new up-regulated genes under drought stress in soybean nodules [J]. Gene, 426: 15-22.

Clough S J, Andrew F B, 1998. Floral dip: a simplified method for Agrobacterium-mediated transformation of *Arabidopsis thaliana* [J]. Plant Journal, 16 (6): 735-743.

Cohen R, Hanan A, Paris H S, 2003. Single-gene resistance to powdery mildew in zucchini squash (*Cucurbita pepo*) [J]. Euphytica, 130: 433-441.

Collinge M, Boller T, 2001. Differential induction of two potato genes, *Stprx2* and *StNAC*, in response to infection by Phytophthora infestans and to wounding [J]. Plant Mol Biol, 46: 521-529.

Consonni C, Humphry M E, Hartmann H A, et al., 2006. Conserved requirement for a plant host cell protein in powdery mildew pathogenesis [J]. Nat Genet, (38): 716-720.

Contin M, Munger H M, 1977. Inheritance of powdery mildew resistance in interspecific crosses with *Cucurbita martinezii* [J]. HortScience, 12: 397.

Craig K L, Tyers M, 1999. The F-box: a new motif for ubiquitin dependent proteolysis in cell cycle regulation and signal transduction [J]. Prog Biophys Mol Biol, 72 (3): 299-328.

Cuzick A, Maguire K, Hammond-Kosack K E, 2009. Lack of the plant signaling component *SGT1b* enhances disease resistance to *Fusarium culmorum* in Arabidopsis buds and flowers [J]. New Phytol, 181: 901-12.

D'Angelo C, Weinl S, Batistic O, et al., 2006. Alternative complex formation of the Ca^{2+}-regulated protein kinase CIPK1 controls abscisic acid dependent and independent stress responses in *Arabidopsis* [J]. Plant J,

48: 857-872.

Dai L, Wang D, Xie X, et al., 2016. The novel gene *VpPR4-1* from *Vitis pseudoreticulata* increases powdery mildew resistance in transgenic *Vitis vinifera* L [J]. Front. Plant Sci., 7: 695.

Das M, Chauhan H, Chhibbar A, et al., 2011. High-efficiency transformation and selective tolerance against biotic and abiotic stress in mulberry, Morus indica cv. K2, by constitutive and inducible expression of tobacco osmotin. Transgenic Res [J]. 20: 231-246.

Datta S K, Muthukrishnan S, 1999. Pathogenesis-related proteins in plants [M]. Boca Raton: CRC Press: 288.

de Azevedo Neto A D, Prisco J T, Eneas-Filho J, et al., 2005. Hydrogen peroxide pretreatment induces salt-stress acclimation in maize plants [J]. J Plant Physiol, 162: 1114-1122.

Delaney T P, Uknes S, Vernooij B, et al., 1994. A central role of salicylic acid in plant disease resistance [J]. Science, 266 (5188): 1247-1250.

Delessert C, Kazan K, Wilson I W, et al., 2005. The transcription factor *ATAF2* represses the expression of pathogenesis-related genes in *Arabidopsis* [J]. Plant Journal, 43 (5): 745-757.

Dharmasiri N, Dharmasiri S, Estelle M, 2005. The F-box protein TIR1 is an auxin receptor [J]. Nature, 435: 441-445.

Dhindsa R S, Plumb-Dhindsa P, Thorpe T A, 1981. Leaf senescence: correlated with increased levels of membrane permeability and lipid peroxidation, and decreased levels of superoxide dismutase and catalase [J]. J Exp Bot, 32: 93-101.

Diatchenko L, Lau Y F, Campbell A P, et al., 1996. Suppression subtraction hybridization: a method for generating differentially regulated or tissue-specific cDNA probes and libraries [J]. Proc Natl Acad Sci USA, 93: 6025-6030.

Dietz K J, Tavakoli N, Kluge C, et al., 2001. Significance of the V-type ATPase for the adaptation to stressful growth conditions and its regulation on the molecular and biochemical level [J]. J Exp Bot, 52: 1969-1980.

Ding W, Song L, Wang X, 2010. Effect of abscisic acid on heat stress tolerance in the calli from two ecotypes of *Phragmites communis* [J]. Biol Plantarum, 54: 607-613.

Ding X, Jiang Y, Hao T, et al., 2016. Effects of heat shock on photosynthetic properties, antioxidant enzyme activity, and downy mildew of cucumber (*cucumis sativus* L.) [J]. Plos One, 11 (4): e0152429.

Ding Y, Sun T, Ao K, et al., 2018. Opposite roles of salicylic acid receptors NPR1 and NPR3/NPR4 in transcriptional regulation of plant immunity [J]. Cell, 6: 173-203.

Ding Z H, Li S M, An X L, et al., 2009. Transgenic expression of *MYB15* confers enhanced sensitivity to abscisic acid and improved drought tolerance in *Arabidopsis thaliana* [J]. J Genet Genomics, 36 (1): 17-29.

Dionisio-Sese M L, Tobita S, 1998. Antioxidant responses of rice seedlings to salinity stress [J]. Plant Sci, 135: 1-9.

Doherty C J, Van Buskirk H A, Myers S J, et al., 2009. Roles for Arabidopsis CAMTA transcription factors in cold-regulated gene expression and freezing tolerance [J]. Plant Cell, 21: 972-984.

Dreher K, Callis J, 2007. Ubiquitin, hormones andbiotic stress in plants [J]. Annals of Botany, 99: 787-822.

Du L, Poovaiah B W, 2005. Ca^{2+}/calmodulin is critical for brassinosteroid biosynthesis and plant growth [J]. Nature, 437: 741-745.

Dubouzet J G, Sakuma Y, Ito Y, et al., 2003. *OsDREB* genes in rice, *Oryza sativa* L., encode transcription activators that function in drought-, high-, salt- and cold-responsive gene expression [J]. Plant J, 33: 751-763.

El Oirdi M, Bouarab K, 2007. Plant signalling components EDS1 and SGT1 enhance disease caused by the necrotrophic pathogen *Botrytis cinerea* [J]. New Phytol, 175: 131-139.

El-Shora H M, 2002. Properties of phenylalanine ammonia-lyase from marrow cotyledons [J]. Plant Sci, 162: 1-7.

Elstner E F, 1982. Oxygen activation and oxygen toxicity [J]. Plant Physiology, 33: 73-96.

Ernst H A, Olsen A N, Larsen S, et al., 2004. Structure of the conserved domain of ANAC, a member of the NAC family of transcription factors [J]. EMBO Rep, 5: 297-303.

Faccioli P, Pecchioni N, Cattivelli L, et al., 2001. Expressed sequence tags from cold-acclimatized barley can identify novel plant genes [J]. Plant Breeding, 120: 497-502.

Fan W, Dong X, 2002. In vivo interaction between NPR1 and transcription factor TGA2 leads to salicylic acid-mediated gene activation in *Arabidopsis*.

Plant Cell, 14: 1377-1389.

Fang Q, Wang Q, Mao H, 2018. AtDIV2, an R-R-type MYB transcription factor of Arabidopsis, negatively regulates salt stress by modulating ABA signaling [J]. Plant Cell Rep. , 37: 1499.

Farquhar G D, Sharkey T D, 1982. Stomatal conductance and photosynthesis [J]. Annual Review Plant Physiology, 33 (1): 317-345.

Feys B, Benedetti C E, Penfold C N, et al. , 1994. *Arabidopsis* mutants selected for resistance to the phytotoxin coronatine are male sterile, insensitive to methyl jasmonate, and resistant to a bacterial pathogen [J]. Plant Cell, 6: 751-759.

Fischer U, Droge-Laser W, 2004. Over-expression of *NtERF5*, a new member of the tobacco Ethylene response transcription factor family enhances resistance to tobacco mosaic virus [J]. Mol Plant Microbe Interact, 17 (10): 1162-1171.

Fleet C M, Sun T P, 2005. A DELLAcate balance: The role of gibberellin in plant morphogenesis [J]. Curr Opin Plant Biol, 8: 77-85.

Fowler S, Thomashow M F, 2002. *Arabidopsis* transcriptome profiling indicated that multiple regulatory pathways are activated during cold acclimation in addition to the CBF cold response pathway [J]. Plant Cell, 14: 1675-1690.

Foyer C H, Descourvieres P, Kunert K J, 1994. Protection against oxygen radicals: an important defense mechanism studied in transgenic plants [J]. Plant Cell and Environment, 17: 507-523.

Fryer M J, Andrews J R, Oxborough k, 1998. Relationship between CO assimilation, photosynthetic electron transport, and active, metabolismin leaves of maize in the field during periods of low temperature [J]. Plant Physiol, 116: 571-580.

Fujita M, Fujita Y, Maruyama K, et al. , 2004. A dehydration-induced NAC protein, *RD26*, is involved in a novel ABA-dependent stress-signaling pathway [J]. Plant J, 39: 863-876.

Fukino N, Yoshioka Y, Sugiyama M, et al. , 2013. Identification and validation of powdery mildew (*Podosphaera xanthii*) -resistant loci in recombinant inbred lines of cucumber (*Cucumis sativus* L.)　[J]. Mol. Breed. , 32: 267-277.

Fukino N, Yoshioka Y, Sugiyama M, et al. , 2013. Identification and validation of powdery mildew (*Podosphaera xanthii*) resistant loci in recombinant inbred lines of cucumber (*Cucumis sativus* L.) [J]. Mol Breeding, 32 (2): 267-77.

Gagne J M, Downes B P, Shiu S H, et al., 2002. The F-box subunit of the SCF E3 complex is encoded by a diverse superfamily of genes in *Arabidopsis* [J]. Proc Natl Acad Sci USA, 99: 11519-11524.

Gamir J, Darwiche R, Van't Hof P, et al., 2017. The sterol-binding activity of pathogenesis-related protein 1 reveals the mode of action of an antimicrobial protein. *Plant J.*, 89: 502-509.

Gao L J, Zhang Y X, 2012. Effects of Salicylic Acid on the Expression of SOD, PPO Isozymes and *NPR1* in Pear [J]. Acta Horticulturae Sinica, 40 (1): 41-48

Gao M, Wang Q, Wan R, et al., 2012. Identification of genes differentially expressed in grapevine associated with resistance to *Elsinoe ampelina* through suppressive subtraction hybridization [J]. Plant Physiol Biochem, 58: 253-268.

Gechev T, Willekens H, Van Montagu M, et al., 2003. Different responses of tobacco antioxidant enzymes to light and chilling stress [J]. J Plant Physiol, 160: 509-515.

Ghassemian M, Lutes J, Chang H S, et al., 2008. Abscisic acid-induced modulation of metabolic and redox control pathways in *Arabidopsis thaliana* [J]. Phytochemistry, 69: 2899-2911.

Ghassemian M, Nambara E, Cutler S, et al., 2000. Regulation of abscisic acid signaling by the ethylene response pathway in *Arabidopsis* [J]. Plant Cell, 12: 1117-1126.

Giannopolitis C N, Ries S K, 1977. Superoxide dismutases: I. Occurrence in higher plants [J]. Plant Physiol, 59: 309-314.

Gibbs G M, Roelants K, O' Bryan M K, 2008. The CAP superfamily: cysteine-rich secretory proteins, antigen 5, and pathogenesis-related 1 proteins-roles in reproduction, cancer, and immune defense [J]. *Endocr. Rev.*, 7: 865-897.

Glazebrook J, 2005. Contrasting mechanisms of defense against biotrophic and necrotrophic pathogens [J]. Annu Rev Phytopathol, 43: 205-227.

Godoy A V, Zanetti M E, San Segundo B, et al., 2001. Identification of a putative *Solanum tuberosum* transcriptional coactivator up-regulated in potato tubers by *Fusarium solani* f. sp. Eumartii infection and wounding [J].

Physiol Plant, 112: 217-222.

Gray W M, del Pozo J C, Walker L, et al., 1999. Identification of an SCF ubiquitin-ligase complex required for auxin response in *Arabidopsis thaliana* [J]. Genes Dev, 13: 1678-1691.

Griffith O W, 1980. Determination of glutathione and glutathione disulfide using glutathione reductase and 2-vinylpyridine [J]. Anal Biochem, 106: 207-212.

Gu Y Q, Wildermuth M C, Chakravarthy S, et al., 2002. Tomato transcription factors Pti4, Pti5, and Pti6 activate defense responses when expressed in *Arabidopsis* [J]. Plant Cell, 14: 817-831.

Gulyani V, Khurana P, 2011. Identification and expression profiling of drought-regulated genes in mulberry (*Morus* sp.) by suppression subtractive hybridization of susceptible and tolerant cultivars [J]. Tree Genet Genomes, 7: 725-738.

Guo C, Yao L, You C, 2016. MID1 plays an important role in response to drought stress during reproductive development [J]. Plant J., 88: 280.

Guo H, Ecker J R, 2003. Plant responses to ethylene gas are mediated by SCF[EBF1/EBF2]-dependent proteolysis of EIN3 transcription factor [J]. Cell, 115: 667-677.

Guo S G, Liu J G, Zheng Y, et al., 2011. Characterization of transcrip-tome dynamics during watermelon fruit development: sequencing, assembly, annotation and gene expression profiles [J]. BMC Genom, 12: 454.

Guo W L, Chen B H, Chen X J, et al., 2018. Transcriptome profiling of pumpkin (*Cucurbita moschata* Duch.) leaves infected with powdery mildew [J]. PloS ONE, 13 (1): e0190175.

Guo W L, Chen B H, Guo Y Y, et al., 2019. Improved Powdery Mildew Resistance of Transgenic *Nicotiana benthamiana* Overexpressing the *Cucurbita moschata CmSGT1* Gene [J]. Front Plant Sci. 10: 955.

Guo W L, Chen R G, Du X H, et al., 2014. Reduced tolerance to abiotic stress in transgenic *Arabidopsis* overexpressing a *Capsicum annuum* multiprotein bridging factor1 [J]. BMC Plant Biol, 14: 138.

Guo W L, Chen R G, Gong Z H, et al., 2012. Exogenous abscisic acid increases antioxidant enzymes and related gene expression in pepper (*Capsicum annuum*) leaves subjected to chilling stress [J]. Genet Mol Res, 11 (4): 4063.

Guo W L, Chen R G, Gong Z H, et al., 2013. Suppression subtractive hybridization analysis of genes regulated by application of exogenous abscisic acid in pepper plant (*Capsicum annuum* L.) leaves under chilling stress [J]. PLoS ONE, 8 (6): e66667.

Guo W L, Wang S B, Chen R G, et al., 2015. Characterization and expression profile of CaNAC2 pepper gene [J]. Front Plant Sci, 6: 755.

Guo Y F, Gan S S, 2006. AtNAP, a NAC family transcription factor, has an important role in leaf senescence [J]. Plant J, 46 (4): 601-612.

Guo Y, Xiong L, Ishitani M, et al., 2002. An *Arabidopsis* mutation in translation elongation factor 2 causes superinduction of CBF/DREB1 transcription factor genes but blocks the induction of their downstream targets under low temperatures [J]. Proc Natl Acad Sci USA, 99 (11): 7786-7791.

Gutierrez L, Mauriat M, Guénin S, et al., 2008. The lack of a systematic validation of reference genes: a serious pitfall undervalued in reverse transcription-polymerase chain reaction (RT-PCR) analysis in plants [J]. Plant Biotechnology Journal, 6: 609-618.

Hakim Ullah, A, Hussain A, Shaban M, et al., 2018. Osmotin: A plant defense tool against biotic and abiotic stresses [J]. Plant Physiol. Biochem., 123: 149-159.

Hamama I B, Yang Z, Gong Q, et al., 2017. *GaMYB85*, an R2R3 MYB gene, in transgenic *Arabidopsis* plays an important role in drought tolerance [J]. BMC Plant Biol., 17: 142.

Hammerschmidt R, Nuckles E M, Kuc J, 1982. Association of enhanced peroxidase activity with induced systemic resistance of cucumber to *Colletotrichum lagenarium* [J]. Physiol Plant Pathol, 20: 73-82.

Han Q Q, Zhang J H, Li H X, et al., 2012. Identification and expression pattern of one stress-responsive *NAC* gene from *Solanum lycopersicum* [J]. Mol Biol Rep, 39: 1713-1720.

Han Z, Xiong D, Schneiter R, et al., 2023. The function of plant PR1 and other members of the CAP protein superfamily in plant-pathogen interactions [J]. Mol. Plant Pathol., 1-18.

Hancock J T, Henson D, Nyirenda M, et al., 2005. Proteomic identification of glyceraldehyde-3-phosphate dehydrogenase as an inhibitory target of hydrogen peroxide in *Arabidopsis* [J]. Plant Physiol Biochem, 43 (9):

828-835.

Hannah M A, Heyer A G, Hincha D K, 2005. A global survey of gene regulation during cold acclimation in *Arabidopsis thaliana* [J]. PLoS Genet, 1- e26.

Hao Y J, Wei W, Song Q X, et al., 2011. Soybean NAC transcription factors promote abiotic stress tolerance and lateral root formation in transgenic plants [J]. Plant Journal, 68: 302-313.

He J, Gu X R, Wei C H, et al., 2016. Identification and Expression Analysis Under Abiotic Stresses of the bHLH Transcription Factor Gene Family in Watermelon [J]. Acta Horticulturae Sinica, 43 (2): 281-294.

He Q, Jones D C, Wei L, et al., 2016. Genome-wide identification of *R2R3-MYB* genes and expression analyses during abiotic stress in *Gossypium raimondii* [J]. Sci. Rep-UK., 6: 22980.

He X J, Mu R L, Cao W H, et al., 2005. *AtNAC2*, a transcription factor downstream of ethylene and auxin signaling pathways, is involved in salt stress response and lateral root development [J]. Plant J, 44: 903-916.

Hegedus D, Yu M, Baldwin D, et al., 2003. Molecular characterization of *Brassica napus* NAC domain transcriptional activators induced in response to biotic and abiotic stress [J]. Plant Mol Biol, 53: 383-397.

Heim M A, Jakoby M, Werber M, et al., 2003. The basic helix-loop-helix transcription factor family in plants: a genome-wide study of protein structure and functional diversity [J]. Mol. Biol. Evol., 20: 735-747.

Hemming M N, Trevaskis B, 2011. Make hay when the sun shines: The role of *MADS-box* genes in temperature-dependant seasonal flowering responses [J]. Plant Sci, 180: 447-453.

Henriques R, JangI C, Chua N H, 2009. Regulated proteolysis in light-related signaling pathways [J]. Current Opinion in Plant Biology, 12 (1): 49-56.

Hermanson O, Glass C K, Rosenfeld M G, 2002. Nuclear receptor coregulators: multiple modes of modification [J]. Trends Endocrinol Metab, 13: 55-60.

Hibara K, Takada S, Tasaka M, 2003. *CUC1* gene activates the expression of SAM related genes to induce adventitious shoot formation [J]. Plant J, 36 (5): 687-696.

Ho M S, Ou C, Chan Y R, et al., 2008. The utility F-box for protein destruction [J]. Cell Mol Life Sci, 65: 1977-2000.

Hodges D M, Lester G E, Munro K D, et al., 2004. Oxidative stress: importance for postharvest quality [J]. HortScience, 39: 924-929.

Hommel M, Khalil-Ahmad Q, Jaimes-Miranda F, et al., 2008. Overexpression of a chimeric gene of the transcriptional co-activator MBF1 fused to the EAR repressor motif causes developmental alteration in *Arabidopsis* and tomato [J]. Plant Sci, 175: 168-77.

Hong S W, Jon J H, Kwak J M, et al., 1997. Identification of a receptor-like protein kinase gene rapidly induced by abscisic acid, dehydration, high salt, and cold treatments in *Arabidopsis thaliana* [J]. Plant Physiol, 113: 1203-1212.

Horváth E, Szalai G, Janda T, 2007. Induction of abiotic stress tolerance by salicylic acid signaling [J]. J Plant Growth Regul, 26: 290-300.

Hoser R, Zurczak M, Lichocha M, et al., 2013. Nucleocytoplasmic partitioning of tobacco N receptor is modulated by SGT1 [J]. New Phytol, 200: 158-171.

Hoshida H, Tanaka Y, Hibino T, et al., 2000. Enhanced tolerance to salt stress in transgenic rice that overexpresses chloroplast glutamine synthetase [J]. Plant Molecular Biology, 43 (1): 103-111.

Hossain Z, Amyot L, McGarvey B, et al., 2012. The translation elongation factor eEF-1Bb1 is involved in cell wall biosynthesis and plant development in *Arabidopsis thaliana* [J]. PLoS ONE, 7 (1): e30425.

Hsieh T H, Lee J T, Yang P T, et al., 2002. Heterology expression of the *Arabidopsis* C-repeat/dehydration response element binding factor 1 gene confers elevated tolerance to chilling and oxidative stresses in transgenic tomato [J]. Plant Physiol, 129: 1086-1094.

Hu H H, Dai M Q, Yao J L, et al., 2006. Overexpressing a NAM, ATAF, and CUC (NAC) transcription factor enhances drought resistance and salt tolerance in rice [J]. Proc Natl Acad Sci USA, 35: 12987-12992.

Hu H, You J, Fang Y, et al., 2008. Characterization of transcription factor gene *SNAC2* conferring cold and salt tolerance in rice [J]. *Plant Mol Biol*, 67: 169-181.

Huang X, Madan A, 1999. CAP3: a DNA sequence assembly program [J]. Genome Res, 9: 868-877.

Hung K T, Kao C H, 2003. Nitric oxide counteracts the senescence of rice leaves induced by abscisic acid [J]. J Plant Physiol, 160: 871-879.

Hung K T, Kao C H, 2004. Hydrogen peroxide is necessary for abscisic acid-induced senescence of rice leaves [J]. J Plant Physiol, 161: 1347-1357.

Iglesias V A, Meins F J, 2000. Movement of plant viruses is delayed in a β-1, 3-glucanase-deficient mutagenized showing a reduced plasmodesmatal size exclusion limit and enhanced callose deposition [J]. Plant J, 21: 157-166.

Ignatova L K, Rudenko N N, Mudrik V A, et al., 2011. Carbonic anhydrase activity in *Arabidopsis thaliana* thylakoid membrane and fragments enriched with PSI or PSII [J]. Photosynth Res, 110: 89-98.

Iqbal M J, Goodwin P H, Leonardos E D, et al., 2012. Spatial and temporal changes in chlorophyll fluorescence images of *Nicotiana benthamiana* leaves following inoculation with *Pseudomonas syringae* pv. *tabaci* [J]. Plant Pathol, 61: 1052-1062.

Ishii H, Fraaije B A, Sugiyama T, et al., 2001. Occurrence and molecular characterization of strobilurin resistance in cucumber powdery mildew and downy mildew [J]. Phytopathology, 91: 1166.

Isutsa D K, Mallowa S O, 2013. Increasing leaf harvest intensity enhances edible leaf vegetable yields and decreases mature fruit yields in multi-purpose pumpkin [J]. Agric Biol Sci, 8: 610-615.

Ito Y, Katsura K, Maruyama K, et al., 2006. Functional analysis of rice DREB1/CBF-type transcription factors involved in cold-responsive gene expression in transgenic rice [J]. Plant Cell Physiol, 47: 141-153.

Jaglo-Ottosen K R, Gilmour S J, Zarka D G, et al., 1998. *Arabidopsis CBF1* overexpression induces *COR* genes and enhances freezing tolerance [J]. Science, 280: 104-106.

Jeong M J, Park S C, Kwon H B, et al., 2000. Isolation and characterization of the gene encoding glyceraldehyde-3-phosphate dehydrogenase [J]. Biochem Bioph Res Co, 278: 192-196.

Jiang M, Zhang J, 2002. Role of abscissic acid in water stress-induced antioxidant defense in leaves of maize seedlings [J]. Free Radic Res, 36: 1001-1015.

Jiang W J, Bai J, Yang X Y, et al., 2012. Exogenous application of abscisic acid, putrescine, or 2, 4-epibrassinolide at appropriate concentrations effectively alleviate damage to tomato seedlings from suboptimal temperature stress [J]. HortTechnology, 22: 137-144.

Jin J, Hewezi T, Baum T J, 2011. The *Arabidopsis* bHLH25 and bHLH27 transcription factors contribute to susceptibility to the cyst nematode Heterodera schachtii [J]. Plant J, 65 (2): 319-328.

Jin Y H, Tao D L, Hao Z Q, 2003. Environmental stresses and redox status of ascorbate [J]. Acta Botanica Sinica, 45: 795- 801.

Jing Y X, Liu J, Liu P, et al., 2019. Overexpression of *TaJAZ1* increases powdery mildew resistance through promoting reactive oxygen species accumulation in bread wheat [J]. Scientific Rep, 9: 5691.

Jones J D G, Dangl J L, 2006. The plant immune system [J]. Nature, 444: 323-329.

Kabe Y, Goto M, Shima D, et al., 1999. The role of human MBF1 as a transcriptional coactivator [J]. J Biol Chem, 274: 34196-34202.

Kang N J, 2009. Induced resistance to powdery mildew by 2, 6-Dichloroisonicotinic acid is associated with activation of active oxygen species-mediated enzymes in cucumber plants [J]. J Japan Soc Hort Sci, 78 (2): 185-194.

Kasuga M, Liu Q, Miura S, et al., 1999. Improving plant drought, salt, and freezing tolerance by gene transfer of a single stress inducible transcription factor [J]. Nat Biotechnol, 17: 287-291.

Kasuga M, Miura S, Shinozaki K, et al., 2004. A combination of the *Arabidopsis DREB1A* gene and stress-inducible RD29A promoter improved drought and low-temperature stress tolerance in tobacco by gene transfer [J]. Plant Cell Physiol, 45: 346-350.

Katagiri T, Takahashi S, Shinozaki K, 2001. Involvement of a novel *Arabidopsis* phospholipase D, AtPLDd, in dehydration-inducible accumulation of phosphatidic acid in stress signaling [J]. Plant J, 26: 595-605.

Kattupalli D, Srinivasan A, Soniya E V, 2021. A genome-wide analysis of pathogenesis-pelated protein-1 (PR-1) genes from *Piper nigrum* reveals its critical role during *Phytophthora capsici* infection [J]. Genes, 12: 1007.

Kepinski S, Leyser O, 2005. The *Arabidopsis* F-box protein TIR1 is an auxin receptor [J]. Nature, 435: 446-451.

Kerdnai-mongkol K, Woodson W R, 1999. Inhibition of catalase by antisense RNA increases susceptibility to oxidative stress and chilling injury in transgenic tomato [J]. J Amer Soc Hort Sci, 124 (4): 330-336.

Kiba A, Nishihara M, NakatsukaT, et al., 2007. Pathogenesis-related protein 1 homologueis an antifungal protein in *Wasabia japonica* leaves and confers resistance to *Botrytis cinerea* in transgenic tobacco. Plant Biotechnol., 24: 247-253.

Kikuchi K, Ueguchi-Tanaka M, Yoshida K T, et al., 2000. Molecular analysis of the NAC gene family in rice [J]. Mol Gen Genet, 262: 1047-1051.

Kim J, Kim H Y, 2006. Molecular characterization of a bHLH transcription factor involved in *Arabidopsis* abscisic acid-mediated response [J]. Biochim Biophys Acta-Gene Struct Expr, 1759 (3-4): 191-194.

Kim K H, Ahn S G, Hwang J H, et al., 2013. Inheritance of resistance to powdery mildew in the watermelon and development of a molecular marker for selecting resistant plants [J]. Hort Environ Biotechnol, 54 (2): 134-140.

Kim M J, Lim G H, Kim E S, et al., 2007. Abiotic and biotic stress tolerance in *Arabidopsis* overexpressing the Multiprotein bridging factor 1a (*MBF1a*) transcriptional coactivator gene [J]. Biochemical and Biophysical Research Communications, 354: 440-446.

Kim S G, Lee A K, Yoon H K, et al., 2008. A membrane-bound NAC transcription factor NTL8 regulates gibberellic acid-mediated salt signaling in *Arabidopsis* seed germination [J]. Plant J, 55: 77-88.

Kim S Y, Kim S G, Kim Y S, et al., 2007. Exploring membrane-associated NAC transcription factors in *Arabidopsis*: implications for membrane biology in genome regulation [J]. Nucleic Acids Res, 35: 203-213.

Kim S, An C S, Hong Y N, et al., 2004. Cold-inducible transcription factor, *CaCBF*, is associated with a homeodomain leucine zipper protein in hot pepper (*Capsicum annuum* L.) [J]. Mol Cell, 18: 300-308.

Kim S, Kang J Y, Cho D I, et al., 2004. ABF2, an ABRE-binding bZIP factor, is an essential component of glucose signaling and its overexpression affects multiple stress tolerance [J]. Plant J, 40: 75-87.

Kim T E, Kim S K, Han T J, et al., 2002. ABA and polyamines act independently in primary leaves of cold-stressed tomato (*Lycopersicon esculentum*) [J]. Physiol Plant, 115 (3): 370-376.

Kim Y S, Kim S G, Park J E, et al., 2006. A membrane-bound NAC transcription factor regulates cell division in *Arabidopsis* [J]. Plant Cell, 18: 3132-3144.

Kipreos E T, Pagano M, 2000. The F-box protein family [J]. Genome

Biology, 1 (5): 1-7.

Kitagawa K, Skowyra D, Elledge S J, et al., 1999. SGT1 encodes an essential component of the yeast kinetochore assembly pathway and a novel subunit of the SCF ubiquitin ligase complex [J]. Mol Cell, 4: 21-33.

Kiyosue T, Yamaguchi-Shinozaki K, Shinozaki K, 1994. *ERD15*, a cDNA for a dehydration-induced gene from *Arabidopsis thaliana* [J]. Plant Physiol, 106: 1707-1707.

Knight H, Zarka D G, Okamoto H, et al., 2004. Abscisic acid induces *CBF* gene transcription and subsequent induction of cold-regulated genes via the CRT promoter element [J]. Plant Physiol, 135: 1710-1717.

Korkmaz A, Korkmaz Y, Demirkiran A R, 2010. Enhancing chilling stress tolerance of pepper seedlings by exogenous application of 5-aminolevulinic acid [J]. Environ Exp Bot, 67: 495-501.

Kornyeyev D, Logan B A, Payton P, et al., 2001. Enhanced photochemical light utilization and decreased chilling-induced photoinhibition of photo system II in cotton overexpressing genes encoding chloroplast-targeted antioxidant enzymes [J]. Physiologia Plantarum, 113 (3): 323-331.

Kosová K, Vítámvás P, Prášil I T, 2011. Expression of dehydrins in wheat and barley under different temperatures [J]. Plant Science, 180 (1): 46-52.

Kovtun Y, Chiu W L, Tena G, et al., 2000. Functional analysis of oxidative stress activated mitogen-activated protein kinase cascade in plants [J]. Proc Natl Acad Sci USA, 97: 2940-2945.

Krause G H, 1994. The role of oxygen in photoinhibition of photosynthesis [M]. in: Foyer C H, Mullineaux P M (Eds.). Causes of photooxidative stress and amelioration of defense systems in plants. Boca Raton. CRC Press: 43-76.

Kreps J A, Wu Y, Chang H S, et al., 2002. Transcriptome changes for *Arabidopsis* in response to salt, osmotic, and cold stress [J]. Plant Physiol, 130: 2129-2141.

Kubo M, Udagawa M, Horiguchi G, et al., 2005. Transcription switches for protoxylem and metaxylem vessel formation [J]. Genes Dev, 19: 1855-1860.

Kuc J, Rush J S, 1985. Phytoalexins [J]. Arch Biochem Biophys, 236: 455-472.

参 考 文 献

Kumar D, Kirti P B, 2015. Pathogen-induced SGT1 of *Arachis diogoi* induces cell death and enhanced disease resistance in tobacco and peanut [J]. Plant Biotechnol J, 13 (1): 73-84.

Kumar S, Kaur G, Nayyar H, 2008. Exogenous application of abscisic acid improves cold tolerance in chickpea (*Cicer arietinum* L.) [J]. J Agron Crop Sci, 194 (6): 449-456.

Kumudini S, Prior E, Omielan J, et al., 2008. Impact of *Phakopsora pachyrhizi* infection on soybean leaf photosynthesis and radiation absorption [J]. Crop Sci, 48: 2343-350.

Kurkela S, Borg-Franck M, 1992. Structure and expression of *KIN2*, one of two cold- and ABA-induced genes of *Arabidopsis thaliana* [J]. Plant Mol Biol, 19: 689-692

Kwon S Y, Jeong Y Z, Lee H S, et al., 2002. Enhanced tolerance of transgenic tobacco plants expressing both superoxide dismutase and ascorbate peroxidase in chloroplasts against methyl viologen-mediated oxidative stress [J]. Plant Cell Environ, 25: 873-882.

Lai Y, Dang F, Lin J, et al., 2013. Overexpression of a Chinese cabbage *BrERF*11 transcription factor enhances disease resistance to *Ralstonia solanacearum* in tobacco [J]. Plant physiology and biochemistry, 62: 70-78.

Laloi C, Apel K, Danon A, 2004. Reactive oxygen signalling: the latest news [J]. Curr Opin Plant Biol, 7: 323-328.

Lang V, Mantyla E, Welin B, et al., 1994. Alterations in water status, endogenous abscisic acid content, and expression of *RAB18* gene during the development of freezing tolerance in *Arabidopsis thaliana* [J]. Plant Physiol, 104: 1341-1349.

Lang V, Palva E T, 1992. The express of a rab-related gene, *rabl8*, is induced by abscisic acid during the cold acclim ation process of *Arabidopsis thaliana* [J]. Plant Mol Biol, 20: 951-962.

Le Hir R, Castelain M, Chakraborti D, et al., 2017. AtbHLH68 transcription factor contributes to the regulation of ABA homeostasis and drought stress tolerance in *Arabidopsis thaliana* [J]. Physiol. Plant, 160 (3): 312-27.

Le Martret B, Poage M, Shiel K, et al., 2011. Tobacco chloroplast transformants expressing genes encoding dehydroascorbate reductase, glutathione reductase, and glutathione-S-transferase, exhibit altered antioxidant metabolism and improved abiotic stress tolerance [J]. Plant

Biotechnol J, 9: 661-673.

Lechner E, Achard P, Vansiri A, et al., 2006. F-box proteins everywhere [J]. Curr Opin Plant Biol, 9: 631-638.

Lee B, Henderson D A, Zhu J K, 2005. The *Arabidopsis* cold-responsive transcriptome and its regulation by ICE1 [J]. The Plant Cell, 17: 3155-3175.

Lee D H, Lee C B, 2000. Chilling stress-induced changes of antioxidant enzymes in the leaves of cucumber: in gel enzyme activity assays [J]. Plant Sci, 159: 75-85.

Lee M H, Lee Y, Hwang I, 2013. In Vivo Localization in *Arabidopsis* Protoplasts and Root Tissue [M]. In: Running M. (eds) G Protein-Coupled Receptor Signaling in Plants. Methods in Molecular Biology (Methods and Protocols), vol 1043. Humana Press, Totowa, NJ.

Lee S, Seo P J, Lee H J, et al., 2012. A NAC transcription factor NTL4 promotes reactive oxygen species production during drought-induced leaf senescence in *Arabidopsis* [J]. Plant J, 70: 831-844.

Lee T M, Lur H S, Chu C, 1993. Role of abscisic acid in chilling tolerance of rice (*Oriza sativa* L.) seedlings. I. Endogenous abscisic acid levels [J]. Plant Cell Environ, 16: 481-490.

Letunic I, Doerks T, Bork P, 2009. SMART 6: recent updates and new developments [J]. Nucleic Acids Research, 37: D229-D232.

Levin J Z, Meyerowttz E M, 1995. UFO: An *Arabidopsis* gene involved in both floral meristem and floral organ development [J]. Plant Cell, 7 (5): 529-548.

Li A, Zhang R, Pan L, et al., 2011. Transcriptome analysis of H_2O_2-treated wheat seedlings reveals a H_2O_2-responsive ratty acid desaturase gene participating in powdery mildew resistance [J]. PLoS One, 12: 1-16.

Li Q, Yu B, Gao Y, et al., 2011. Cinnamic acid pretreatment mitigates chilling stress of cucumber leaves through altering antioxidant enzyme activity [J]. J Plant Physiol, 168: 927-934.

Li R, Zhang L L, Yang X M, et al., 2019. Transcriptome analysis reveals pathways facilitating the growth of tobacco powdery mildew in arabidopsis [J]. Phytopathology Research, 1: 7.

Li S, Wang Z, Tang B, et al., 2021. A pathogenesis-related protein-like gene is involved in the *Panax notoginseng* defense response to the root rot pathogen [J]. Front. Plant Sci., 11: 610176.

Li W, Li M, Zhang W, et al., 2004. The plasma membrane-bound phospholipase Dd enhances freezing tolerance in *Arabidopsis thaliana* [J]. Nat Biotechnol, 22: 427-433.

Li W, Qi L, Lin X, et al., 2009. The expression of manganese superoxide dismutase gene from *Nelumbo nucifera* responds strongly to chilling and oxidative stresses [J]. J Integr Plant Biol, 51: 279-286.

Li X Y, Gao L, Zhang W H, et al., 2015. Characteristic expression of wheat *PR5* gene in response to infection by the leaf rust pathogen, *Puccinia triticina* [J]. Plant Interact., 10: 132-141.

Li Y, Liu Y, Zhang J G, 2010. Advances in the research on the AsA-GSH cycle in horticultural crops [J]. Front Agric China, 4: 84-90.

Li Z T, Dhekney S A, Gray D J, 2011. *PR-1* gene family of grapevine: A uniquely duplicated *PR-1* gene from a Vitis interspecific hybrid confers high level resistance to bacterial disease in transgenic tobacco [J]. Plant Cell Rep., 30 (1): 1-11.

Li Z, Wang S, Tao Q, et al., 2012. A putative positive feedback regulation mechanism in CsACS2 expression suggests a modified model for sex determination in cucumber (*Cucumis sativus* L.) [J]. J Exp Bot, 63: 4475-4484.

Lichtenthaler H K, 1987. Chlorophylls and carotenoids: pigments of photosynthetic biomemranes [J]. Methods Enzymol, 148: 350-382.

Lima A L S, DaMatta F M, Pinheiro H A, et al., 2002. Photochemical responses and oxidative stress in two clones of *Coffea canephora* under water deficit conditions [J]. Environ Exp Bot, 47: 239-247.

Liu L, White M J, MacRae T H, 1999. Transcription factors and their genes in higher plants: functional domains, evolution and regulation [J]. Eur J Biochem, 262: 247-257.

Liu Q X, Jindra M, Ueda H, et al., 2003. Drosophila MBF1 is a co-activator for Tracheae Defective and contributes to the formation of tracheal and nervous systems [J]. Development, 130: 719-728.

Liu Q, Kasuga M, Sakuma Y, et al., 1998. Two transcription factors, DREB1 and DREB2, with an EREBP/AP2 DNA binding domain separate two cellular signal transduction pathways in drought- and low-temperature-responsive gene expression, respectively, in *Arabidopsis* [J]. Plant Cell, 10: 1391-1406.

Liu W Y, Chiou S J, Ko C Y, et al., 2011. Functional characterization of three Ethylene response factor genes from *Bupleurum kaoi* indicates that *BkERFs* mediate resistance Botrytis cinerea [J]. Journal of Plant Physiology, 168: 375-381.

Liu X Q, Bai X Q, Qian Q, et al., 2005. OsWRKY03, a rice transcriptional activator that functions in defense signaling pathway upstream of *OsNPR1* [J]. Cell Research, 15 (8): 593-603.

Liu Y, Jiang H, Zhao Z, et al., 2011. Abscisic acid is involved in brassinosteroids-induced chilling tolerance in the suspension cultured cells from *Chorispora bungeana* [J]. J Plant Physiol, 168: 853-862.

Liu Y, Schiff M, Dinesh-Kumar S P, 2002. Virus-induced gene silencing in tomato [J]. Plant J, 31: 777-786.

Liu Z J, Guo Y K, Bai J G, 2010. Exogenous hydrogen peroxide changes antioxidant enzyme activity and protects ultrastructure in leaves of two cucumber ecotypes under osmotic stress [J]. J Plant Growth Regul, 29: 171-183.

Liu Z Q, Liu Y Y, Shi L P, et al., 2016. SGT1 is required in PcINF1/SRC2-1 induced pepper defense response by interacting with SRC2-1 [J]. Scientific Reports, (6): 21651.

Logan B A, Grace S C, Adams W W, et al., 1998. Seasonal differences in xanthophyll cycle characteristics and antioxidants in *Mahonia repens* growing in different light environments [J]. Oecologia, 116: 9-17.

Loik M E, Nobe P S, 1993. Exogenous abscisic acid mimics cold acclimation for cacti differing in freezing tolerance [J]. Plant Physiol, 103: 871-876.

Lu H C, Lin J H, Chua A C N, et al., 2012. Cloning and expression of pathogenesis-related protein 4 from jelly fig (*Ficus awkeotsang* Makino) achenes associated with ribonuclease, chitinase and anti-fungal activities [J]. Plant Physiol. Bioch., 56, 1-13.

Lu P L, Chen N Z, An R, et al., 2007. A novel drought-inducible gene, *ATAF1*, encodes a NAC family protein that negatively regulates the expression of stress-responsive genes in *Arabidopsis* [J]. Plant Mol Biol, 63 (2): 289-305.

Lu X, Yang L, Yu M, et al., 2017. A novel *Zea mays* ssp. *mexicana* L. MYC-type *ICE*-like transcription factor gene *ZmmICE1*, enhances freezing tolerance in transgenic *Arabidopsis thaliana* [J]. Plant Physiol

参考文献

Biochem, 113: 78-88.

Lyons J M, Raison J K, 1970. Oxidative activity of mitochondria isolated from plant tissues sensitive and resistant to chilling injury [J]. Plant physiol, 45: 386-389.

Lyzenga W J, Stone S L, 2011. Protein ubiquitination: an emerging theme in plant abiotic stress tolerance [J]. Am J Plant Sci Biotechnol, 5: 1-11.

Ma X W, Ma F W, Mi Y F, et al., 2008. Morphological and physiological responses of two contrasting malus species to exogenous abscisic acid application [J]. Plant Growth Regul, 56: 77-87.

Magbanua Z V, Moraes C M, Brooks T D, et al., 2007. Is catalase activity one of the factors associated with maize resistance to *Aspergillus flavus*? [J]. Mol Plant-Microbe Interact, 20: 697-706.

Mahajan S, Tuteja N, 2005. Cold, salinity and drought stresses: An overview [J]. Archives of Biochemistry and Biophysics, 444: 139-158.

Maldonado-Calderón M T, Sepúlveda-García E, Rocha-Sosa M, 2012. Characterization of novel F-box proteins in plants induced by biotic and abiotic stress [J]. Plant Science, 185: 208-217.

Malinovsky F G, Batoux M, Schwessinger B, et al., 2014. Antagonistic regulation of growth and immunity by the arabidopsis basic helix-loop-helix transcription factor homolog of brassinosteroid enhanced expression2 interacting with increased leaf inclination1 binding bhlh1 [J]. Plant Physiol, 164 (3), 1443-1455.

Manghwar H, Hussain A, 2022. Mechanism of tobacco osmotin gene in plant responses to biotic and abiotic stress tolerance: A brief history [J]. Biocell., 46 (3): 623.

Maraschin S F, Caspers M, Potokina E, et al., 2006. cDNA array analysis of stress-induced gene expression in barley androgenesis [J]. Physiol Plant, 127: 535-550.

Marino D, Froidure S, Canonne J, et al., 2019. Arabidopsis ubiquitin ligase MIEL1 mediates degradation of the transcription factor MYB30 weakening plant defence [J]. Nat. Commun., 4: 1476.

Mateos R M, Bonilla-Valverde D, del Río L A, et al., 2008. NADP-dehydrogenases from pepper fruits: effect of maturation [J]. Physiol Plantarum, 135 (2): 130-139.

Matsui A, Ishida J, Morosawa T, et al., 2008. *Arabidopsis* transcriptome analysis under drought, cold, high-salinity and ABA treatment conditions using a tiling array [J]. Plant Cell Physiol, 49: 1135-1149.

Mauro M F D, Iglesias M J, Arce D P, et al., 2012. MBF1s regulate ABA-dependent germination of *Arabidopsis* seeds [J]. Plant Signal Behav, 7 (2): 188-192.

MccreightJ D, 2003. Genes for resistance to powdery mildew races 1 and 2U. S. in melon PI 313970 [J]. Hort. Sci., 38: 591-594.

McKenna N J, O'Malley B W, 2002. Combinatorial control of gene expression by nuclear receptors and coregulator [J]. Cell, 108: 465-474.

Mckersie B D, Bowley S R, Jones K S, 1999. Winter survival of transgenic alfalfa overexpressing superoxide dismutase [J]. Plant Physiology, 119: 839-848.

Meldau S, Baldwin I T, Wu J, 2011. For security and stability: SGT1 in plant defense and development [J]. Plant Signal Behav, 6: 1479-1482.

Mengist T, 2012. Plant immunity to necrotrophs [J]. Annu. Rev. Phytopathol. 50: 267-294.

Mittelheuser C J, van Steveninck R F M, 1969. Stomatal closure and inhibition of transpiration induced by (RS) -abscisic acid [J]. Nature, 221: 281-282.

Mittler R, 2002. Oxidative stress, antioxidants and stress tolerance [J]. Trends Plant Sci, 7: 405-410.

Moon J, Parry G, Estelle M, 2004. The ubiquitin-proteasome pathway and plant development [J]. Plant Cell, 16: 3181-3195.

Morishita T, Kojima Y, Maruta T, et al., 2009. *Arabidopsis* NAC transcription factor, *ANAC078*, regulates flavonoid biosynthesis under high-light [J]. Plant Cell Physiology, 50: 2210-2222.

Moura J C M S, Bonine C A V, Viana J O F, et al., 2010. Abiotic and biotic stresses and changes in the lignin content and composition in plants [J]. J Integr Plant Biol, 52 (4): 360-376.

Mukherjee S P, Choudhuri M A, 1983. Implications of water stress-induced changes in the levels of endogenous ascorbic acid and hydrogen peroxide in vigna seedlings [J]. Physiol Plant, 58: 166-170.

Mukhopadhyay A, Vij S, Tyagi A K, 2004. Overexpression of a zinc-finger protein gene from rice confers tolerance to cold, dehydration, and salt stress in transgenic tobacco [J]. Proc Natl Acad Sci USA, 101: 6309-6314.

Muskett P, Parker J, 2003. Role of SGT1 in the regulation of plant R gene signaling [J]. Microbes Infect, 5: 969-976

Nakano Y, Asada K, 1981. Hydrogen peroxide is scavenged by ascorbate-specific peroxidase in spinach chloroplasts [J]. Plant Cell Physiol, 22: 867-880.

Nakashima K, Tran L P, Nguyen D V, et al., 2007. Functional analysis of a NAC-type transcription factor *OsNAC6* involved in abiotic and biotic stress-responsive gene expression in rice [J]. Plant J, 51: 617-630.

Nanasato Y, Tabei Y, 2015. Cucumber (*Cucumis sativus* L.) and Kabocha Squash (*Cucurbita moschata* Duch). (eds) Agrobacterium Protocols [J]. Methods in Molecular Biology, 1223: 299-310.

Nanasato Y, Konagaya K, Okuzaki A, et al., 2011. *Agrobacterium*-mediated transformation of kabocha squash (*Cucurbita moschata* Duch) induced by wounding with aluminum borate whiskers [J]. Plant Cell Rep, 30 (8): 1455-1464.

Nanasato Y, Konagaya K, Okuzaki A, 2011. Agrobacterium-mediated transformation of kabocha squash (*Cucurbita moschata* Duch) induced by wounding with aluminum borate whiskers [J]. Plant Cell Rep., 30: 1455.

Nanasato Y, Tabei Y, 2015. Cucumber (*Cucumis sativus* L.) and Kabocha Squash (*Cucurbita moschata* Duch). (eds) Agrobacterium Protocols [J]. Methods in Molecular Biology, 1223: 299-310.

Naoki T, HirofumiK, TakashiK, 2004. Expression and interaction analysis of *Arabidopsis Skpl2* related genes [J]. Plant and Cell Physiology, 45 (1): 83-91.

Nayyar H, Bains T S, Kumar S, 2005. Chilling stressed chickpea seedlings: effect of cold acclimation, calcium and abscisic acid on cryoprotective solutes and oxidative damage [J]. Environ Exp Bot, 54: 275-285.

Neill S J, Desikan R, Clarke A, et al., 2002. Hydrogen peroxide and nitric oxide as signalling molecules in plants [J]. J Exp Bot, 53: 1237-1247.

Ngadze E, Icishahayo D, Coutinho T A, et al., 2012. Role of polyphenol oxidase, peroxidase, phenylalanine ammonialyase, chlorogenic acid, and total soluble phenols in resistance of potatoes to soft rot [J]. Plant Dis, 96: 186-192.

Nguyen H T, Leipner J, Stamp P, et al., 2009. Low temperature stress in maize (*Zea mays* L.) induces genes involved in photosynthesis and signal transduction as studied by suppression subtractive hybridization [J]. Plant

Physiol Biochem, 47: 116-122.

Ning X, Wang X, Gao X, et al., 2014. Inheritances and location of powdery mildew resistance gene in melon Edisto47 [J]. Euphytica, 195: 34353.

Noël L D, Cagna G, Stuttmann J, et al., 2007. Interaction between SGT1 and cytosolic/nuclear HSC70 chaperones regulates *Arabidopsis* immune responses [J]. The Plant Cell, 19 (12): 4061-4076.

Nylander M, Svensson J, Palva E T, et al., 2001. Stress-induced accumulation and tissue-specific localization of dehydrins in *Arabidopsis thaliana* [J]. Plant Mol Biol, 45: 263-279.

Oberschall A, Deak M, Torok K, et al., 2000. A novel aldose/aldehyde reductase protects transgenic plants against lipid peroxidation under chemical and drought stresses [J]. Plant J, 24: 437-446.

Oh S K, Lee S, Yu S H, et al., 2005. Expression of a novel NAC domain-containing transcription factor (*CaNAC1*) is preferentially associated with incompatible interactions between chili pepper and pathogens [J]. Planta, 222: 876-887.

Ohnishi T, Sugahara S, Yamada T, et al., 2005. *OsNAC6*, a member of the NAC gene family, is induced by various stresses in rice [J]. Genes Genet Syst, 80: 135-139.

Olsen A N, Ernst H A, Leggio L L, et al., 2005. NAC transcription factors: structurally distinct, functionally diverse [J]. Trends Plant Sci, 10: 79-87.

Onate-Sanchez L, Anderson J P, Young J, et al., 2007. *AtERF14*, a member of the *ERF* family of transcription factors, plays a nonredundant role in plant defense [J]. Plant Physiol, 143 (1): 400-409.

Orvar B L, Sangwan V, Omann F, et al., 2000. Early steps in cold sensing by plant cells: the role of actin cytoskeleton and membrane fluidity [J]. Plant J, 23: 785-794.

Osakabe Y, Maruyama K, Seki M, et al., 2005. Leucine-rich repeat receptor-like kinase1 is a key membrane-bound regulator of abscisic acid early signaling in *Arabidopsis* [J]. Plant Cell, 17: 1105-1119.

Pan I C, Li C W, Su R C, et al., Ectopic expression of an EAR motif deletion mutant of *SlERF3* enhances tolerance to salt stress and *Ralstonia solanacearum* in tomato [J]. Planta, 2010, 232: 1075-1086.

Paquis S, Mazeyrat-Gourbeyre F, Fernandez O, et al., 2011. Characterization of an F-box gene up-regulated by phytohormones and upon biotic and abiotic

stresses in grapevine [J]. Mol Biol Rep, 38: 3327-3337.
Paris H S, Cohen R, 2002. powdery mildew-resistant summer squash hybrids having higher yields than their susceptible, commercial counterparts [J]. Euphytica, 124: 121-128.
Park J M, Park C J, Lee S B, et al., 2001. Over-expression of the tobacco Tsi1 gene encod-ing an*EREBP/AP2* type transcription factor enhances resistance against pathogen attack and osmotic stress in tobacco [J]. Plant Cell, 13: 1035-1046.
Park J M, Park C J, Lee S B, et al., 2001. Over-expression of the tobacco Tsi1 gene encoding an *EREBP/AP2*-type transcription factor enhances resistance against pathogen attack and osmotic stress in tobacco [J]. Plant Cell, 13: 1035-1046.
Park M R, Yun K Y, Mohanty B, et al., 2010. Supra-optimal expression of the cold-regulated OsMyb4 transcription factor in transgenic rice changes the complexity of transcriptional network with major effects on stress tolerance and panicle development [J]. Plant Cell Environ, 33: 2209-2230
Pawłowski T A, 2009. Proteome analysis of Norway maple (*Acer platanoides* L.) seeds dormancy breaking and germination: influence of abscisic and gibberellic acids [J]. BMC Plant Biology, 9: 48.
Peart J R, Lu R, Sadanandom A, et al., 2002. Ubiquitin ligase-associated protein SGT1 is required for host and nonhost disease resistance in plants. Proceedings of the National Academy of Sciences, 99 (16): 10865-10869.
Peng H, Yu X, Cheng H, et al., 2010. Cloning and characterization of a novel NAC family gene *CarNAC1* from chickpea (*Cicer arietinum* L.) [J]. Molecular Biotechnology, 44: 30-40.
Perez-Garcia A, Romero D, Fernandez-Ortuno D, et al., 2009. The powdery mildew fungus *Podosphaera fusca* (synonym *Podosphaera xanthii*), a constant threat to cucurbits [J]. Mol Plant Pathol, 10: 153-60.
Pérez-García A, Romero D, Fernández-Ortuño D, et al., 2009. The powdery mildew fungus Podosphaera fusca (synonym *Podosphaera xanthii*), a constant threat to cucurbits [J]. Mol. Plant Pathol. 10: 153-160.
Perez-Prat E, Narasimhan M L, Binzel M L, et al., 1992. Induction of a putative Ca^{2+}-ATPase messenger-RNA in NaCl-adapted cells [J]. Plant Physiol, 100: 1471-1478.

Perl A, Perl-Treves R, Galili S, et al., 1993. Enhanced oxidative-stress defense in transgenic potato expressing tomato Cu/Zn superoxide dismutases [J]. Theor Appl Genet, 85: 568-576.

Petroski M D, Deshaies R J, 2005. Function and regulation of cullin-RINGubiquitin ligases [J]. Nature Reviews Molecular Cell Biology, 6 (1): 9-20.

Petrov V D, VAN Breusegem F, 2012. Hydrogen peroxide: A central hub for information flow in plant cell [J]. AoB Plants, 2012: pls014.

Pieterse C M J, Vander Does D, Zamioudis C, et al., 2012. Hormonal modulation of plant immunity [J]. Annu Rev Cell Dev Biol, 28 (1): 489-521.

Polanco L R, Rodrigues F A, Nascimento K J T, et al., 2014. Photosynthetic gas exchange and antioxidative system in common bean plants infected by *Colletotrichum lindemuthianum* and supplied with silicon [J]. Trop Plant Pathol, 39 (1): 35-42.

Potuschak T, Lechner E, Parmentier Y, et al., 2003. EIN3-dependent regulation of plant ethylene hormone signaling by two *Arabidopsis* F-box proteins: EBF1 and EBF2 [J]. Cell, 115: 679-689.

Prasad T K, 1997. Role of catalase in inducing chilling tolerance in pre-emergent maize seedlings [J]. Plant Physiol, 114: 1369-1376.

Pressman E, Shaked R, Firon N, 2006. Exposing pepper plants to high day temperatures prevents the adverse low night temperature symptoms [J]. Physiol Plantarum, 126: 618-626.

Puranik S, Bahadur R P, Srivastava P S, et al., 2011. Molecular cloning and characterization of a membrane associated NAC Family gene, *SiNAC* from Foxtail Millet [*Setaria italica* (L.) P. Beauv.] [J]. Mol Biotechnol, 49: 138-150.

Pushpanathan M, Rajendhran J, Jayashree S, et al., 2012. Identification of a novel antifungal peptide with chitin-binding property from marine metagenome [J]. Protein Peptide Lett, 19: 1289-1296.

Qi X H, Xu Q, Shen L P, et al., 2010. Identification of differentially expressed genes between powdery mildew resistant near-isogenic line and susceptible line of cucumber by suppression subtractive hybridization [J]. Scientia Horticulturae, 126 (1): 27-32.

Rabbani M A, Maruyama K, Abe H, et al., 2003. Monitoring expression profiles of rice genes under cold, drought, and high-salinity stresses and abscisic acid application using cDNA microarray and RNA gel-blot analyses [J]. Plant Physiol, 133: 1755-1767.

Riechmann J L, Ratcliffe O J, 2000. A genomic perspective on plant transcription factors, Curr. Opin. Plant Biol., 3: 423.

Rigoyen M L, Garceau D C, Bohorquez-Chaux A, et al., 2020. Genome-wide analyses of cassava Pathogenesis-related (PR) gene families reveal core transcriptome responses to whitefly infestation, salicylic acid and jasmonic acid [J]. BMC Genom., 21 (1): 93.

Risseeuw E P, Daskalchuk T E, Banks T W, et al., 2003. Protein interaction analysis of SCF ubiquitin E3 ligase subunits from *Arabidopsis* [J]. Plant Journal, 34 (6): 753-767.

Rizhsky I, Liang H, Mittler R, 2002. The combined effect of drought stress and heat shock on gene expression in tobacco [J]. Plant Physiol, 130: 1143-1151.

Robert-Seilaniantz A, Grant M, Jones J D G, 2011. Hormone crosstalk in plant disease and defense: more than just jasmonate-salicylate antagonism [J]. Annu Rev Phytopathol, 49: 317-343.

Rolfe S A, Scholes J D, 2010. Chlorophyll fluorescence imaging of plant pathogen interactions [J]. Protoplasma, 247: 163-175.

Saibo N J M, Lourenc O T, Oliveira M M, 2009. Transcription factors and regulation of photosynthetic and related metabolism under environmental stresses [J]. Annals of Botany, 103: 609-623.

Sanders D, Pelloux J, Brownlee C, et al., 2002. Calcium at the crossroads of signaling [J]. Plant Cell, 14: S401-S417.

Sangwan V, Foulds I, Singh J, et al., 2001. Cold-activation of Brassica napus BN115 promoter is mediated by structural changes in membranes and cytoskeleton, and requires Ca^{2+} in flux [J]. Plant J, 27: 1-12.

Sato Y, Murakami T, Funatsuki H, et al., 2001. Heat shock-mediated APX gene expression and protection against chilling injury in rice seedlings [J]. Journal of Experimental Botany, 52 (354): 145-151.

Scandalios J G, 2011. Wheat resistance to leaf blast mediated by silicon [J]. Australas Plant Pathol, 40: 28-38.

Schaedle M, 1977. Chloroplast glutathione reductase [J]. Plant Physiol, 59: 1011-1012.

Scharte J, Schön H, Weis E, 2005. Photosynthesis and carbohydrate metabolism in tobacco leaves during an incompatible interaction with *Phytophthora nicotianae* [J]. Plant Cell Environ, 28: 1421-1435.

Schiøtt M, Palmgren MG, 2005. Two plant Ca^{2+} pumps expressed in stomatal guard cells show opposite expression patterns during cold stress [J]. Physiol Plant, 124: 278-283.

Schultz J, Milpetz F, Bork P, et al., 1998. SMART, a simple modular architecture research tool: identification of signalling domains [J]. Proceedings of the National Academy of Sciences USA, 95: 5857-5864.

Schwechheimer C, Willige B C, Zourelidou M, et al., 2009. Examining protein stabilityand its relevance for plant growth and development [J]. Methods in Molecular Biology, 479: 147-171.

Scofield S R, Huang L, Brandt A S, et al., 2005. Development of a virus-induced gene-silencing system for hexaploid wheat and its use in functional analysis of the *Lr21*-mediated leaf rust resistance pathway [J]. Plant Physiol, 138: 2165-2173.

Seki M, Ishida J, Narusaka M, et al., 2002. Monitoring the expression pattern of around 7,000 *Arabidopsis* genes under ABA treatments using a full-length cDNA microarray [J]. Funct Integr Genomics, 2: 282-291.

Seki M, Kamei A, Yamaguchi-Shinozaki K, et al., 2003. Molecular responses to drought, salinity and frost: common and different paths for plant protection [J]. Curr Opin Biotech, 14: 194-199.

Seki M, Narusaka M, Ishida J, et al., 2002. Monitoring the expression profiles of 7000 *Arabidopsis* genes under drought, cold and high salinity stresses using a full-length cDNA microarray [J]. Plant J, 31: 279-292.

Selote D S, Khanna-Chopra R, 2006. Drought acclimation confers oxidative stress tolerance by inducing co-ordinated antioxidant defense at cellular and subcellular level in leaves of wheat seedlings [J]. Physiol Plant, 127: 494-506.

Seo J S, Joo J, Kim M J, et al., 2011. OsbHLH148, a basic helix-loop-helix protein, interacts with OsJAZ proteins in a jasmonate signaling pathway leading to drought tolerance in rice [J]. Plant J, 65 (6): 907-21.

Seo P J, Kim M J, Park J Y, et al., 2010. Cold activation of a plasma membrane-tethered NAC transcription factor induces a pathogen resistance response in *Arabidopsis* [J]. Plant J, 61: 661-671.

参 考 文 献

Seo P J, Kim S G, Park C M, 2008. Membrane-bound transcription factors in plants [J]. Trends Plant Sci, 13: 550-556.

Seung Y Y, Hwang B K, 1996. Differential induction and accumulation of β-1, 3-glucanase and chitinase isoforms in soybean hypocotyls and leaves after compatible and incompatible infection with *Phytophthora megasperma* f. sp. *glycinea* [J]. Physiol Mol Plant Pathol, 48: 179-192.

Shan C, Liang Z, 2010. Jasmonic acid regulates ascorbate and glutathione metabolism in *Agropyron cristatum* leaves under water stress [J]. Plant Sci, 178: 130-139.

Shen Q H, Zhou F, Bieri S, et al., 2003. Recognition specificity and RAR1/SGT1 dependency in barley *Mla* disease resistance alleles to the powdery mildew fungus [J]. Plant Cell, 15: 732-744.

Shi Y, Tian S, Hou L, et al., 2012. Ethylene signaling negatively regulates freezing tolerance by repressing expression of CBF and type-A ARR genes in *Arabidopsis* [J]. Plant Cell, 24: 2578-2595.

Shin D, Moon S J, Han S, et al., 2011. Expression of StMYB1R-1, a novel potato single myb-like domain transcription factor, increases drought tolerance [J]. Plant Physiol., 155: 421.

Shin S H, Pak J H, Kim M J, et al., 2014. An acidic Pathogenesis-Related1 gene of *Oryza grandiglumis* is involved in disease resistance response against bacterial infection [J]. Plant Pathol. J., 30 (2): 208-214.

Shinozaki K, Yamaguchi-Shinozaki K, Seki M, 2003. Regulatory network of gene expression in the drought and cold stress responses [J]. Curr Opin Plant Biol, 6: 410-417.

Siddique Z, Akhtar K P, Hameed A, et al., 2014. Biochemical alterations in leaves of resistant and susceptible cotton genotypes infected systemically by cotton leaf curl Burewala virus [J]. J Plant Interact, 9: 702-711.

Slooten L, Capian K, Van-Camp W, 1995. Factors affecting the enhancement of oxidative stress tolerance in transgenic tobacco overexpressing manganese superoxide dismutase in the chloroplasts [J]. Plant Physiology, 107 (3): 737-750.

Smalle J, Vierstra R D, 2004. The ubiquitin 26S proteasome proteolytic pathway [J]. Annu Rev Plant Biol, 55: 555-590.

Smallwood M, Bowles D J, 2002. Plants in a cold climate [J]. Philos Trans R Soc Lond B Biol Sci, 357: 831-846.

Soliman A M, Idriss M H, El-Meniawi F A, et al., 2019. Induction of Pathogenesis-Related (PR) proteins as a plant defense mechanism for controlling the cotton whitefly bemisia tabaci [J]. Alex. J. Agric. Sci., 64 (2): 107-122.

Souer E, van Houwelingen A, Kloos D, et al., 1996. The no apical meristem gene of Petunia is required for pattern formation in embryos and flowers and is expressed at meristem and primordia boundaries [J]. Cell, 85 (2): 159-170.

Souza P F N, Silva F D A, Carvalho F E L, et al., 2017. Photosynthetic and biochemical mechanisms of an ems-mutagenized cowpea associated with its resistance to cowpea severe mosaic virus [J]. Plant Cell Rep, 36: 219-234.

Sperotto R A, Ricachenevsky F K, Duarte G L, et al., 2009. Identification of up-regulated genes in flag leaves during rice grain filling and characterization of *OsNAC5*, a new ABA-dependent transcription factor [J]. Planta, 230 (5): 985-1002.

Spoel S H, Dong X, 2012. How do plants achieve immunity? Defence without specialized immune cells. Nat [J]. Rev. Immunol, 12: 89-100.

Steponkus P L, 1984. Role of the plasma membrane in freezing injury and cold acclimation [J]. Annual Review of Plant Physiology, 35: 543-584.

Stevens R, Page D, Gouble B, et al., 2008. Tomato fruit ascorbic acid content is linked with monodehydroascorbate reductase activity and tolerance to chilling stress [J]. Plant Cell Environ, 31: 1086-1096.

Sticher L, Mauch-Mani B, Metraux J P, 1998. Systemic acquired resistance [J]. Annu Rev. Phytopathol, 35: 235-270.

Stirnberg P, van De Sande K, Leyser H M, 2002. MAX1 and MAX2 control shoot lateral branching in *Arabidopsis* [J]. Development, 129: 1131-1141.

Strader L C, Ritchie S, Soule J D, et al., 2004. Recessive-interfering mutations in the gibberellin signaling gene SLEEPY1 are rescued by overexpression of its homologue, SNEEZY [J]. Proc Natl Acad Sci USA, 101: 12771-12776.

Su C F, Wang Y C, Hsieh T H, et al., 2010. A novel MYBS3-dependent pathway confers cold tolerance in rice [J]. Plant Physiol, 153: 145-158.

Sun H, Wu S, Zhang G, et al., 2017. Karyotype stability and unbiased fractionation in the paleo-allotetraploid Cucurbita genomes [J]. Mol Plant, 10 (10): 1293

Sun X, Lei T, Yuan S, et al., 2005. Progress in research of dehydrins [J]. Journal of Wuhan Botanical Research, 23 (3): 299-304.

Suzuki N, Koussevitzky S, Mittler R, et al., 2011. ROS and redox signaling in the response of plants to abiotic stress [J]. Plant Cell Environ, 35: 259-270.

Suzuki N, Rizhsky L, Liang H, et al., 2005. Enhanced tolerance to environmental stress in transgenic plants expressing the transcriptional coactivator multiprotein bridging factor 1c [J]. Plant Physiol, 139: 1313-1322.

Suzuki N, Sejima H, Tam R, et al., 2011. Identification of the MBF1 heat-response regulon of *Arabidopsis thaliana* [J]. The Plant Journal, 66: 844-851.

Takemaru K I, Harashima S, Ueda H, et al., 1998. Yeast coactivatior MBF1 mediates GCN4-dependent transcriptional activation [J]. Mol Cell Biol, 18: 4971-4976.

Takemaru K I, Li F Q, Ueda H, et al., 1997. Multiprotein bridging factor 1 (MBF1) is an evolutionarily conserved transcriptional coactivator that connects a regulatory factor and TATA element-binding protein [J]. Proc Natl Acad Sci USA, 94: 7251-7256.

Tan X, Zheng N, 2009. Hormone signaling through protein destruction: a lesson from plants [J]. Am J Physiol Endocrinol Metab, 296: E223-E227.

Tanaka Y, Sano T, Tamaoki M, et al., 2005. Ethylene inhibits abscisic acid-induced stomatal closure in *Arabidopsis* [J]. Plant Physiology, 138: 2337-2343.

Tang L, Kwon S Y, Kim S H, et al., 2006. Enhanced tolerance of transgenic potato plants expressing both superoxide dismutase and ascorbate peroxidase in chloroplasts against oxidative stress and high temperature [J]. Plant Cell Rep, 25: 1380-1386.

Tang Y M, Liu M Y, Gao S Q, et al., 2012. Molecular characterization of novel TaNAC genes in wheat and overexpression of *TaNAC2a* confers drought tolerance in tobacco [J]. Physiologia Plantarum, 144: 210-224.

Tang Y, Liu Q, Liu Y, et al., 2017. Overexpression of *NtPR-Q* upregulates multiple defense-related genes in *Nicotiana tabacum* and enhances plant resistance to *Ralstonia solanacearum* [J]. Front. Plant Sci., 8: 1963.

Tatagiba S D, Neves F W, Bitti A L F E, et al., 2016. Changes in gas exchange and antioxidant metabolism on rice leaves infected by *Monographella albescens* [J]. Trop Plant Pathol, 41: 33-41.

Teige M, Scheikl E, Eulgem T, et al., 2004. The MKK2 pathway mediates cold and salt stress signaling in *Arabidopsis* [J]. Mol Cell, 15: 141-152.

Teper-Bamnolker P, Samach A, 2005. The flowering integrator FT regulates SEPAL-LATA3 and fruitfull accumulation in *Arabidopsis* leaves [J]. Plant Cell, 17: 2661-2675.

Thomas C E, Levi A, Caniglia E, 2005. Evaluation of U. S. plant introductions of watermelon for resistance to powdery mildew [J]. Hortscience A Publication of the American Society for Horticultural Science, 40 (1): 154-156.

Thomas C L, Jones L, Baulcombe D C, et al., 2001. Size constraints for targeting post- transcriptional gene silencing and for RNA-directed methylation in *Nictiana benthamuana* using a potato virus X vector [J]. Plant J, 24: 417-425.

Thomas S G, Sun T P, 2004. Update on gibberellin signaling. A tale of the tall and the short [J]. Plant Physiol, 135: 668-676.

Thomashow M F, 1999. Plant cold acclimation: Freezing tolerance genes and regulatory mechanisms [J]. Annu Rev Plant Physiol Plant Mol Biol, 50: 571-599.

Thomma B P, Eggermont K, Broekaert W F, et al., 2000. Disease development of several fungi on *Arabidopsis* can be reduced by reatment with methy jasmonate [J]. Plant Physiology and Biochemistry, 38 (5): 421-427.

Thompson J D, Higgins D G, Gibson T J, 1994. CLUSTAL W: Improving the sensitivity of progressive multiple sequence alignment through sequence weighting, position-specific gap penalties and weight matrix choice [J]. Nucleic Acids Res, 22: 4673-4680.

Tiwari S B, Hagen G, Guilfoyle T J, 2004. AUX/IAA proteins contain a potent transcriptional repression domain [J]. Plant Cell, 16: 533-543.

Tojo T, Tsuda K, Yoshizumi T, et al., 2009. *Arabidopsis* MBF1s control leaf cell cycle and its expansion [J]. Plant Cell Physiol, 50: 254-64.

Toledo-Ortiz G, Huq E, Quail P H, 2003. The *Arabidopsis* basic/helixloop-helix transcription factor family [J]. Plant Cell, 15: 1749-1770.

Topuz A, Ozdemir F, 2007. Assessment of carotenoids, capsaicinoids and ascorbic acid composition of some selected pepper cultivars (*Capsicum annuum* L.) grown in Turkey [J]. J Food Compos Anal, 20: 596-602.

Torres M A, 2010. ROS in biotic interactions [J]. Physiol Plant, 138: 414-429.

Tran L S, Nakashima K, Sakuma Y, et al., 2004. Isolation and functional analysis of *Arabidopsis* stress-inducible NAC transcription factors that bind to a drought-responsive cis-element in the early responsive to dehydration stress 1 promoter [J]. Plant Cell, 16: 2481-2498.

Tran L S, Quach T N, Guttikonda S K, et al., 2009. Molecular characterization of stress-inducible *GmNAC* genes in soybean [J]. Mol Genet Genomics, 281: 647-664.

Tsuda K, Tsuji T, Hirose S, et al., 2004. Three *Arabidopsis* MBF1 homologs with distinct expression profiles play roles as transcriptional co-activator [J]. Plant Cell Physiol, 45: 225-231.

Tsuda K, Yamazaki K, 2004. Structure and expression analysis of three subtypes of *Arabidopsis* MBF1 genes [J]. Biochim Biophys Acta, 1680: 1-10.

Ulmasov T, Hagen G, Guilfoyle T J, 1999. Activation and repression of transcription by auxin-response factors [J]. Proc Natl Acad Sci USA, 96: 5844-5849.

Uppalapati S R, Ishiga Y, Ryu C M, et al., 2010. *SGT1* contributes to coronatine signaling and *Pseudomonas syringae* pv. *tomato* disease symptom development in tomato and Arabidopsis [J]. New Phytol, 189: 83-93.

Van Loon L C, Rep M, and Pieterse C M, 2006. Significance of inducible defense related proteins in infected plants [J]. Annu. Rev. Phytopathol., 44: 135-162.

van Loon L, Bakker P A H M, Pieterse C M J, 1998. Systemic resistance duced by rhizos phere bacteria [J]. Annu Rew Phytopathol, 36: 453-483.

Van Schie C C N, Takken F L W, 2014. Susceptibility genes 101: how to be a good host [J]. Annu. Rev. Phytopathol., 52: 551-581.

Vaultier M N, Cantrel C, Vergnolle C, et al., 2006. Desaturase mutants reveal that membrane rigidification acts as a cold perception mechanism upstream of the diacylglycerol kinase pathway in *Arabidopsis* cells [J]. FEBS Lett, 580: 4218-4223.

Verslues P E, Zhu J K, 2005. Before and beyond ABA: upstream sensing and internal signals that determine ABA accumulation and response under abiotic stress [J]. Biochem Soc Trans, 33: 375-379.

Vij S, Tyagi A K, 2007. Emerging trends in the functional genomics of the abiotic stress response in crop plants [J]. Plant Biotechnol J, 5: 361-380.

Vinocur B, Altman A, 2005. Recent advances in engineering plant tolerance to abiotic stress: achievements and limitations [J]. Current Opinion in Biotechnology, 16: 123-132.

Vogel J T, Zarka D G, Van Buskirk H A, et al., 2005. Roles of the CBF2 and ZAT12 transcription factors in configuring the low temperature transcriptome of *Arabidopsis* [J]. Plant J, 41: 195-211.

Wada H, Gombos Z, Murata N, 1994. Contribution of membrane-lipids to the ability of the photosynthetic machinery to tolerate temperature stress [J]. Proc Natl Acad Sci USA, 91 (10): 4273-4277.

Wan H, Yuan W, Ruan M, et al., 2011. Identification of reference genes for reverse transcription quantitative real-time PCR normalization in pepper (*Capsicum annuum* L.) [J]. Biochem. Biophys. Res Commun, 416: 24-30.

Wang F, Lin R, Feng J, et al., 2015. Wheat bHLH transcription factor gene, *TabHLH060*, enhances susceptibility of transgenic *Arabidopsis thaliana* to *Pseudomonas syringae* [J]. Physiol Mol Plant Pathol, 90: 123-30.

Wang F, Shen S, Zhao C, et al., 2022. TaPR1 interacts with TaTLP1 *via* the αIV helix to be involved in wheat defense to *Puccinia triticina* through the CAPE1 motif [J]. Front Plant Sci., 26: 874654.

Wang F, Yuan S, Wu W, et al., 2020. TaTLP1 interacts with TaPR1 to contribute to wheat defense responses to leaf rust fungus [J]. PLoS Genet., 16 (7): e1008713.

Wang J E, Li D W, Gong Z H, et al., 2013. Optimization of virus-induced gene silencing in pepper (*Capsicum annuum* L.) [J]. Genetics and Molecular Research, 12 (3): 2492-2506.

Wang J E, Li D W, Zhang Y L, et al., 2013. Defence responses of pepper (*Capsicum annuum* L.) infected with incompatible and compatible strains of *Phytophthora capsici* [J]. Eur. J. Plant Pathol., 136: 625-638.

Wang K, Uppalapati S R, Zhu X, et al., 2010. SGT1 positively regulates the process of plant cell death during both compatible and incompatible plant-pathogen interactions [J]. Mol Plant Pathol, 11: 597-611.

参 考 文 献

Wang P, Liu J C, Zhao Q Y, 2002. Studies on nutrient composition and utilization of pumpkin fruit [J]. Inner Mongolia Agric Univ, 23: 52-54.

Wang W, Vinocur B, Altman A, 2003. Plant responses to drought, salinity and extremetemperatures: Towards genetic engineering for stress tolerance [J]. Planta, 218: 1-14.

Wang X, Li G, Gao X, et al., 2011. Powdery mildew resistance gene (*Pm-AN*) located in a segregation distortion region of melon LGV [J]. Euphytica, 180 (3): 421-428.

Wang Y L, Ma F W, Li M J, et al., 2011. Physiological responses of kiwifruit plants to exogenous ABA under drought conditions [J]. Plant Growth Regul, 64: 63-74.

Wang Z, Xiao Y, Chen W, et al., 2010. Increased vitamin C content accompanied by an enhanced recycling pathway confers oxidative stress tolerance in *Arabidopsis* [J]. J Integr Plant Biol, 52: 400-409.

Wasternack C, Hause B, 2013. Jasmonates: biosynthesis, perception, signal transduction and action in plant stress response, growth and development. An update to the 2007 review in annals of botany [J]. Ann Bot, 111 (6): 1021-58.

Wei K F, Chen H Q, 2018. Comparative functional genomics analysis of bHLH gene family in rice, maize and wheat [J]. BMC Plant Biol, 18: 309.

Wimmers L E, Ewing N N, Bennett A B, 1992. Higher-plant Ca^{2+}-ATPase-primary structure and regulation of messenger-RNA abundance by salt [J]. Proc Natl Acad Sci U S A, 89: 9205-9209.

Woo H R, Chung K M, Park J H, et al., 2001. ORE9, an F-box protein that regulates leaf senescence in *Arabidopsis* [J]. Plant Cell, 13: 1779-1790.

Wright M, 1974. The effect of chilling on ethylene production, membrane permeability, and water loss of leaves of *Phaseolus vulgaris* [J]. Planta, 120: 63-69.

Wu H, Ye H, Yao R, et al., 2015. OsJAZ9 acts as a transcriptional regulator in jasmonate signaling and modulates salt stress tolerance in rice [J]. Plant Sci, 232: 1-12.

Wu T Q, Luo S B, Wang R, et al., 2014. The first Illumina-based de novo transcriptome seqencingand analysis of pumpkin (*Cucurbita moschata*

Duch.) and SSR marker development [J]. Mol Breeding, 34: 1437-1447.

Wu T, Cao J, 2010. Molecular cloning and expression of a bush related CmV1, gene in tropical pumpkin [J]. Mol Biol Rep, 37 (2): 649.

Wu Y, Deng Z, Lai J, et al. , 2009. Dual function of *Arabidopsis* ATAF1 in abiotic and biotic stress responses [J]. Cell Res, 19: 1279-1290.

Xiang J, Li X L, Yin L, et al. , 2017. A candidate RxLR effector from *Plasmopara viticola* can elicit immune responses in *Nicotiana benthamiana* [J]. BMC Plant Biol. , 17: 75.

Xie D X, Feys B F, James S, et al. , 1998. *COI1*: An *Arabidopsis* gene required for jasmonate-regulated defense and fertility [J]. Science, 280: 1091-1094.

Xie Q, Frugis G, Colgan D, et al. , 2000. *Arabidopsis* NAC1 transduces auxin signal downstream of TIR1 to promote lateral root development [J]. Genes Dev, 14: 3024-3036.

Xie Q, Sanz-Burgos A P, Guo H, et al. , 1999. GRAB proteins, novel members of the NAC domain family, isolated by their interaction with a geminivirus protein [J]. Plant Mol Biol, 39: 647- 656.

Xing L, Qian C, Cao A, et al. , 2013. The *Hv-SGT1* gene from *Haynaldia villosa* contribute to resistances towards both biotrophic and hemi-biotrophic pathogens in common wheat (*Triticum aestivum* L.) [J]. PloS One, 8: e72571.

Xiong L, Zhu J K, 2003. Regulation of abscisic acid biosynthesis [J]. Plant Physiol, 133: 29-36.

Xu J, Zhang Y X, Guan Z Q, et al. , 2008. Expression and function of two dehydrins under environmental stressed in *Brassica juncea* L. [J]. Mol Breeding, 21: 431-438.

Xu P, Jiang L, Wu J, et al. , 2014. Isolation and characterization of a pathogenesis-related protein 10 gene (*GmPR10*) with induced expression in soybean (*Glycine max*) during infection with *Phytophthora sojae* [J]. Mol. Biol. Rep. , 41: 4899-4909.

Xu W, Jiao Y, Li R, et al. , 2014. Chinese wild-growing Vitis amurensis ICE1 and ICE2 encode MYC-type bHLH transcription activators that regulate cold tolerance in *Arabidopsis* [J]. PLoS One, 9 (7): e102303.

Xu Z S S, Chen M C, Li L C C, et al. , 2008. Functions of the *ERF* transcription factor in plants [J]. Botany-botanique, 86 (9), 969-977.

Xue-Xuan X, Hong-Bo S, Yuan-Yuan M, et al. , 2010. Biotechnological implications from abscisic acid (ABA) roles in cold stress and leaf senescence as animportant signal for improving plant sustainable survival under abiotic-stressed conditions [J]. Crit Rev Biotechnol, 30: 222-230.

Yadav M, Jain S, Tomar R, et al., 2010. Medicinal and biological potential of pumpkin: an updated review [J]. Nutr Res Rev, 23: 184-190.

Yamaguchi M, Kubo M, Fukuda H, et al., 2008. VASCULAR-RELATED NAC-DONAIN7 is involved in differentiation of all types of xylem vessels in *Arabidopsis* roots and shoots [J]. Plant J, 55 (4): 652-664.

Yamaguchi-Shinozaki K, Shinozaki K, 1993. The plant hormone abscisic acid mediates the drought-induced expression but not the seed-specific expression of *RD22*, a gene responsive to dehydration stress in *Arabidopsis thaliana* [J]. Mol Gen Genet, 238: 17-25.

Yamamura C, Mizutani E, Okada K, et al., 2015. Diterpenoid phytoalexin factor, a bHLH transcription factor, plays a central role in the biosynthesis of diterpenoid phytoalexins in rice [J]. Plant J, 84 (6): 1100-13.

Yan Q, Cui X, Su L, et al., 2014. GmSGT1 is differently required for soybean Rps genes-mediated and basal resistance to *Phytophthora sojae* [J]. Plant Cell Rep., 33: 1275.

Yan Y S, Chen X Y, Yang K, et al., 2011. Overexpression of an F-box protein gene reduces abiotic stress tolerance and promotes root growth in rice [J]. Molecular Plant, 4 (1): 190-197.

Yang L, Ji W, Zhu Y, et al., 2010. GsCBRLK, a calcium/calmodulin-binding receptor-like kinase, is a positive regulator of plant tolerance to salt and ABA stress [J]. J Exp Bot, 61 (9): 2519-2533.

Yang L, Wang C C, Guo W D, et al., 2006. Differential expression of cell wall related genes in the elongation zone of rice roots under water deficit [J]. Rus J Plant Physiol, 53: 390-395.

Yang R C, Deng C T, Ouyang B, et al., 2011. Molecular analysis of two salt-responsive NAC-family genes and their expression analysis in tomato [J]. Mol Biol Rep, 38: 857-863.

Yang Y W, Wu Y, Pirrello J, et al., 2010. Silencing *Sl-EBF1* and *Sl-EBF2* expression causes constitutive ethylene response phenotype, accelerated plant senescence, and fruit ripening in tomato [J]. Journal of Experimental Botany, 61 (3): 697-708.

Yanhui C, Xiaoyuan Y, Kun H, et al., 2006. The MYB transcription factor superfamily of Arabidopsis: expression analysis and phylogenetic comparison with the rice MYB family [J]. Plant Mol. Biol., 60: 107.

Yao L F, Yang B, Xian B S, et al., 2020. The R2R3-MYB transcription factor BnaMYB111L from rapeseed modulates reactive oxygen species accumulation and hypersensitive-like cell death [J]. Plant Physiol. Bioch., 147: 280.

Yoshimura K, Masuda A, Kuwano M, et al., 2008. Programmed proteome response for drought avoidance/tolerance in the root of a C-3 xerophyte (wild watermelon) under water deficits [J]. Plant Cell Physiol, 49: 226-241.

Yu H, Wu J, Xu N F, et al., 2007. Roles of F-box proteins in plant hormone responses [J]. Acta Biochimica et Biophysica Sinica, 39 (12): 915-922.

Yu S, Zhang X, Guan Q, et al., 2007. Expression of a carbonic anhydrase gene is induced by environmental stresses in rice (*Oryza sativa* L.) [J]. Biotechnol Lett, 29: 89-94.

Yu W Y, Pan Y G, Ke Y Q, et al., 2006. stuides on photosynthesis of sweet potato under the stress of sweet potato scab [J]. Chinese Journal of Eco-Agriculture, 14 (4): 161-164.

Yu Y, Guo D, Li G, et al., 2019. The grapevine R2R3-type MYB transcription factor VdMYB1 positively regulates defense responses by activating the stilbene synthase gene 2 (VdSTS2) [J]. BMC Plant Biol., 19: 478.

Zanetti M E, Chan R L, Godoy A V, et al., 2004. Homeodomain-leucine zipper proteins interact with a plant homologue of the transcriptional co-activator multiprotein bridging factor 1 [J]. J Biochem Mol Biol, 37: 320-324.

Zhai H, Bai X, Zhu Y, et al., 2010. A single-repeat R3-MYB transcription factor MYBC1 negatively regulates freezing tolerance in *Arabidopsis* [J]. Biochem Biophys Res Commun, 394: 1018-1023.

Zhang G, Chen M, Li L C, et al., 2009. Over-expression of the soybean *GmERF3* gene, an *AP2/ERF* type transcription factor for increased tolerances to salt, drought, and diseases in transgenic tobacco [J]. Journal of Experimental Botany, 60 (13): 3781-3796.

Zhang H B, Zhang D B, Chen J, et al., 2004. Tomato stress-responsive factor*TSRF*1 interacts with Ethylene responsive element GCC box and regulates pathogen resistance to *Ralstonia solanacearum* [J]. Plant

Molecular Biology, 55: 825-834.

Zhang N, Schulman A B, Song L Z, 2002. Structure of the Cull Rbxl-Skpl-F-box Skp2 SCF ubiquitin ligase complex [J]. Nature, 146: 703-709.

Zhang W, Jiang B, Li W, et al., 2009. Polyamines enhance chilling tolerance of cucumber (*Cucumis sativus* L.) through modulating antioxidative system [J]. Sci Hortic, 122: 200-208.

Zhang W, Wang C, Qin C, et al., 2003. The oleate-stimulated phospholipase D, PLDd, and phosphatidic acid decrease H_2O_2-induced cell death in *Arabidopsis* [J]. Plant Cell, 15: 2285-2295.

Zhang W, Yu L, Zhang Y, et al., 2005. Phospholipase D in the signaling networks of plant response to abscisic acid and reactive oxygen species [J]. Biochim Biophys Acta, 1736: 1-9.

Zhang X H, Rao X L, Shi H T, et al., 2011. Overexpression of a cytosolic glyceraldehyde-3-phosphate dehydrogenase gene *OsGAPC3* confers salt tolerance in rice [J]. Plant Cell Tiss Organ Cult, 107: 1-11.

Zhang X, Chen L, Shi Q, et al., 2019. *SlMYB102*, an R2R3-type MYB gene, confers salt tolerance in transgenic tomato, Plant Sci., 291: 110356.

Zhang Y, Tang H R, Luo Y, et al., 2009. Responses of antioxidant enzymes and compounds in strawberry (Fragaria x ananassa 'Toyonaka') to cold stress [J]. New Zeal J Crop Hort, 37: 383-390.

Zhang Y, Xu W, Li Z, et al., 2008. F-Box protein *DOR* functions as a novel inhibitory factor for abscisic acid-induced stomatal closure under drought stress in *Arabidopsis* [J]. Plant Physiol, 148: 2121-2133.

Zhang Y, Yin X, Xiao Y, et al., 2018. An ethylene response factor-myb transcription complex regulates furaneol biosynthesis by activating quinone oxidoreductase expression in strawberry [J]. Plant physiol., 178: 189.

Zhao C, Avci U, Grant E H, et al., 2008. XND1, a member of the NAC domain family in *Arabidopsis thaliana*, negatively regulates lignocellulose synthesis and programmed cell death in xylem [J]. Plant J, 53 (3): 425-436.

Zhao Q, Xiang X, Liu D, et al., 2018. Tobacco Transcription Factor NtbHLH123 Confers Tolerance to Cold Stress by Regulating the NtCBF Pathway and Reactive Oxygen Species Homeostasis. Front [J]. Plant Sci, 9: 381.

Zhao X C, Schaller G E, 2004. Effect of salt and osmotic stress upon expression of the ethylene receptor ETR1 in *Arabidopsis thaliana* [J]. Febs Letters, 562: 189-192.

Zhong R Q, Demura T, Ye Z H, 2006. SND1, a NAC domain transcription factor, is a key regulator of secondary wall synthesis in fibers of *Arabidopsis* [J]. Plant Cell, 18: 3158-3170.

Zhou B Y, Guo Z F, Lin L, 2006. Effects of abscisic acid application on photosynthesis and photochemistry of *Stylosanthes guianensis* under chilling stress [J]. Plant Growth Regul, 48: 195-199.

Zhou B Y, Guo Z F, Liu Z L, 2005. Effects of abscisic acid on antioxidant systems of *Stylosanthes guianensis* (Aublet) Sw. under chilling stress [J]. Crop Sci, 45: 599-605.

Zhou J G, Hu H L, Li X Z, et al., 2010. Identification of a resource of powdery mildew resistance in *Cucurbita moschata* [J]. Acta Horticulturae, 871: 141-146.

Zhou Q Y, Tian A G, Zou H F, et al., 2008. Soybean WRKY-type transcription factor genes, *GmWRKY13*, *GmWRKY21*, and *GmWRKY54*, confer differential tolerance to abiotic stresses in transgenic *Arabidopsis* plants [J]. Plant Biotechnol J, 6: 486-503.

Zhu J H, Dong C H, Zhu J K, 2007. Interplay between cold-responsive gene regulation, metabolism and RNA processing during plant cold acclimation [J]. Curr Opin Plant Biol, 10: 290-295.

Zhu W, Lu M H, Gong Z H, et al., 2011. Cloning and expression of a small heat shock protein gene *CaHSP24* from pepper under abiotic stress [J]. African Journal of Biotechnology, 10 (25): 4968-49.

Zhu X, Xiao K, Cui H, et al., 2017. Overexpression of the *Prunus sogdiana* NBS-LRR subgroup gene *PsoRPM2* promotes resistance to the root-knot nematode meloidogyne incognita in tobacco [J]. Front Microbiol, 8: 2113.

图书在版编目（CIP）数据

南瓜和辣椒抗逆生理代谢及抗性基因挖掘 / 郭卫丽著. -- 北京：中国农业出版社，2024.10
ISBN 978-7-109-31220-3

Ⅰ.①南… Ⅱ.①郭… Ⅲ.①南瓜－抗性－研究②辣椒－抗性－研究 Ⅳ.①S642.134②S641.334

中国国家版本馆 CIP 数据核字（2023）第 197102 号

中国农业出版社出版
地址：北京市朝阳区麦子店街 18 号楼
邮编：100125
责任编辑：国 圆 谢志新
版式设计：杨 婧 责任校对：张雯婷
印刷：中农印务有限公司
版次：2024 年 10 月第 1 版
印次：2024 年 10 月北京第 1 次印刷
发行：新华书店北京发行所
开本：880mm×1230mm 1/32
印张：9.75
字数：270 千字
定价：68.00 元

版权所有·侵权必究
凡购买本社图书，如有印装质量问题，我社负责调换。
服务电话：010-59195115　010-59194918